SOTHIS PRESS

Sothis Press aims to completely reimagine the public domain book with an emphasis on universal access to captivatingly designed print editions. We believe that technology will enable every single book ever published to be perpetually accessible in beautiful print form. Sothis is beauty in ubiquity.

www.sothis.press
1 Waterhouse Square, London, ECIN 2ST

What Is Electricity?

John Trowbridge

SOTHIS PRESS

ISBN 978-1-0365-0106-8

This edition is published by
Sothis Press © 2023. The company
is registered in England and Wales.
The Company Number is 14900389.

THE INTERNATIONAL SCIENTIFIC SERIES

WHAT IS ELECTRICITY?

BY

JOHN TROWBRIDGE, S. D.

RUMFORD PROFESSOR AND LECTURER ON THE APPLICATIONS OF SCIENCE
TO THE USEFUL ARTS, HARVARD UNIVERSITY, AND DIRECTOR
OF THE JEFFERSON PHYSICAL LABORATORY

ILLUSTRATED

NEW YORK
D. APPLETON AND COMPANY
1896

PREFACE.

I AM often asked the question, "What is electricity?" and I have endeavoured in this book to give in a popular manner the present views of scientific men in regard to this question. According to modern ideas, the continuance of all life on the earth is due to the electrical energy which we receive from the sun; and physics, in general, can be defined as that subject which treats of the transformations of energy. I have therefore presented the varied phenomena of electricity in such a manner that the reader can perceive the physicist's reasons for supposing that all space is filled with a medium which transmits electro-magnetic waves to us from the sun.

In Tyndall's Heat as a Mode of Motion, Tait's Recent Advances in Physical Science, and Stewart's Conservation of Energy, the relations between work done and heat produced have been treated in a popular manner; but I am not acquainted with any treatise in which Maxwell's great generalization, entitled the Electro-Magnetic Theory of Light, has been made the basis of a popular treatment. The wide-embracing nature of this theory can be seen when we realize that, according to it, all phenomena of light, heat as well as those of electricity, are manifestations of electrical energy.

I have used in this treatise various popular lectures which I have delivered, and certain articles which I have published in the Chautauquan, the Popular Science Monthly, the American Journal of Science, and the London Philosophical Magazine.

I realize fully the difficulty of stating accurately in a popular exposition what is more definitely expressed in mathematical language. If I have succeeded in giving the general reader an idea of the present direction of investigation in the science of electricity I shall consider myself fortunate.

CONTENTS.

WHAT IS ELECTRICITY?

CHAPTER I.

THE STANDPOINT OF PHYSICISTS IN REGARD TO THE QUESTION, "WHAT IS ELECTRICITY?"

THE question, "What is electricity?" is often asked as if a short and lucid answer could be given which a liberally educated person could comprehend. In order to understand the grounds upon which a natural philosopher bases his attempts to answer this question one must consider the entire field of activity in which we find ourselves—a field which we shall see is now believed to be due to the electrical energy of the sun. The subject of physics can be said to be the study of the transformations of energy, and it is the object of this treatise to describe in a popular manner how a great intellectual movement which is now going on silently and steadily, constantly pushing back the limit of our ignorance and occasionally lifting its veil, has taken the place of unsystematic investigation and sterile philosophic vagaries.

Indeed, the characteristic of physical science to-day is its reliance upon patient observation and the study of the transformation of electricity into light and heat or the transformation of heat into electricity. A well-trained physicist listens with as much intolerance to the speculations of a philosopher on the origin of force as Moltke would have listened to the philosophical faculty of the University of Berlin on the origin of war.

The great modern intellectual movement in physical science resides in the abandonment of mere speculation, and the substitution for it of the study of detail and the investigation of the economy of Nature in the transformations of the store of energy which has been vouchsafed to the world. Every university in the world now has its systematic laboratories; and the methods of patient investigation which characterize the laboratory study of science are slowly creeping into the study of other subjects, notably that of law, and are destined, we believe, to be adopted in all subjects.

Lord Salisbury, in an address before the British Association for the Advancement of Science at Oxford, 1894, said: "Science in the universities for many generations bore a signification different from that which belongs to it in this assembly. It represented the knowledge which alone in the Middle Ages was thought worthy of the name of science. It was the knowledge gained not by external observation, but by mere reflection. The student's microscope was turned inward upon the recesses of his own brain, and when the supply of facts and realities failed, as it very speedily did, the scientific imagination was not wanting to furnish to successive generations an interminable series of conflicting speculations."

The chief characteristic of this marked intellectual method, the most noteworthy movement which scientific education has seen, is the accurate study of the transformations of energy; for we perceive that there is much to occupy the investigator in this subject and that constant work is repaid by results, whereas philosophical speculations upon what force is and what electricity is gives us no vantage ground over the metaphysicians.

Agnosticism in physical science is a hopeful creed when it is enlivened by a quick imagination, which is employed by the laboratory worker to suggest clues to follow in his study of the transformations of energy.

There is no tendency to restrain the imagination in this attitude of scientific agnosticism. The physicist of to-day has his ethers and his atoms just as the ancient Greek and Roman philosophers had theirs, and he pictures to himself invisible motions far more subtle than entered the imagination of Aristotle or Democritus. The natural philosopher of to-day, however, differs in this essential respect from the ancient philosopher: he measures. If his heat measures do not agree with his hypotheses of vortical or atomic motions, he rejects his attractive hypotheses instead of hugging them.

In my attempt, therefore, to give in a popular manner the speculations of physicists upon the question, "What is electricity?" we must carefully bear in mind the standpoint of investigators and the way in which they hold their hypotheses. However attractive the hypotheses, they are ruthlessly abandoned as soon as the touchstone, the measurement of the heat equivalent of the motion, is not satisfied by the hypotheses. It is not often that one finds an intelligent appreciation of this attitude of holding hypotheses in suspense which is characteristic of the best minds in science of to-day, and indeed the task set for the physicists is not fully comprehended. They are a small body of men to whom the world looks especially for exact information. By means of patient measurement in the great field of the transformations of energy they have been able to supply the world with the most exact, if not perfectly exact, information which it now possesses. Since the subject of this treatise is a popular presentation of what I regard as the real subject of physics—the transformations of energy—and of the greatest generalization in that subject (Maxwell's Theory of Electro-magnetic Origin of Light and Heat), I can not do better than to quote his words in regard to the attitude of mind and mental characteristics of the order of men whose speculations we are about to study:

"With respect to the 'material sciences,' they appear to me to be the appointed road to all *scientific* truth, whether metaphysical, mental, or social. The knowledge which exists on these subjects derives a great part of its value from ideas suggested by analogies from the material sciences, and the remaining part, though valuable and important to mankind, is not *scientific*, but aphoristic. The chief *philosophical* value of physics is that it gives the mind something distinct to lay hold of, which, if you don't, Nature at once tells you you are wrong. Now, every stage of this conquest of truth leaves a more or less presentable trace on the memory, so that materials are furnished here more than anywhere else for the investigation of the great question, 'How does knowledge come?'

"I have observed that the practical cultivators of science (e. g., Sir J. Herschel, Faraday, Ampère, Oersted, Newton, Young,), although differing excessively in turn of mind, have all a distinctness and a freedom from the tyranny of words in dealing with questions of order, law, etc., which pure speculators and literary men never attain." *

I have said that the measurements of heat produced by motion constitute our great touchstone in testing the truth of physical hypotheses. Tyndall's treatise on Heat a Mode of Motion was an epoch-making book in popular estimation. It certainly brought readers to realization of the great truth that all motion has its equivalent in heat. The treatise was called Heat considered as a Mode of Motion. However much we may be inclined to criticise the title—for we can not properly class the motions which produce heat in a class as modes—we must recognise that it presented certain aspects of the law of conservation of energy in a singularly lucid manner. There is no hint, however, in the treatise of a possible relation between motion, light, heat, and electricity. The lectures which form the basis of Tyndall's treatise were delivered before the Royal Institution in 1862, and in the index to the volume one finds no reference to Maxwell and his great electromagnetic theory of light. In fact, Maxwell was ready to come forward with his great generalization at the time Tyndall was bringing the world to a realizing sense of the conservation of energy.

Maxwell's theory that light and heat are phenomena of electro-magnetic waves which come to us from the sun is now the greatest generalization in physical science, and in stating it Maxwell lighted a torch which has illumined many hitherto dark regions. It is safe to affirm that the entire world is now working upon this great hypothesis. According to the electro-magnetic theory of

---

* Letter to R. B. Litchfield, Life of James Clerk Maxwell (Campbell and Garnett) p. 305.

light, the only difference between light, heat, and electricity consists in the length of waves in the ether of space. The sun is the source of electro-magnetic waves, and the earth is the scene of transformations of electric energy. A piece of coal burning in a grate has therefore a long electro-magnetic history. It owed its origin to electro-magnetic waves, and in burning it gives out again electro-magnetic waves, of which we can only detect the light and heat manifestations.

The Rumford professorship in Harvard University was endowed by Count Rumford in order to promote our knowledge of light and heat; for many years the lectures given under the endowment were devoted exclusively to the subject of light and heat and the conservation of energy—a subject to which Count Rumford's experiments may be said to have given one of the primal impulses. The lecturer and student to-day, however, is compelled to approach the subject of light and heat through the broadest study of electromagnetism.

Tyndall, I have said, in his Heat considered as a Mode of Motion, has happily illustrated the great doctrine of the conservation of energy. No truth is better established in physics than that of the equivalence between work and the heat produced, and the modern developments of electricity afford means for richly illustrating it. The claims of the great founders of this hypothesis have been set forth by Tyndall in his treatise, and by Tait in his Recent Advances in Physical Science. My object in this treatise is to call attention to the transformations of energy rather than to discuss the claims of priority; to follow the protean forms under which energy manifests itself, rather than to measure the equivalence of these forms of energy.

If we wish at this stage of our presentation of the subject of the transformation of energy exultantly to proclaim the results of the methods of the measurements of heat equivalents of motion, we can point to the commercial applications of electricity, which are all based upon exact experiments on the equivalence between the heat produced by the current and the work done by the steam engine which produces the current.

The doctrine of the conservation of energy is little more than a hundred years old. Count Rumford made his celebrated measurement of the heat produced by the boring of a cannon in 1789. Yet in this time the world, working with this powerful theory, has made greater progress than it did with its physical speculations during the comparatively immense historic period which preceded 1789.

Not only in the commercial world do we find that the great physical doctrine has accomplished great results. In the subject of medicine the doctrine is daily being recognised; for the application of external heat to the human body to diminish the effort of the human organism to supply heat, or to supplement this effort in very young children or in very old people, is now clearly understood. The death of very old persons at night is often due to the want of heat. They can be said to freeze to death. Very young children often perish because sufficient heat is not supplied to them. It seems, therefore, to be both physically and physiologically unphilosophical to expose the limbs of young children to the cold air. In hospitals, during severe operations, the patient is placed upon a warm table and is afterward surrounded with hot-water bags or similar contrivances to supply heat. This external heat facilitates the various transformations of energy which are going on in the human organism, and too great a demand is not made upon the internal mechanism to supply this heat.

The study of the transformations of energy in the human body is a far more difficult one than the study of the duty of an ordinary steam engine, for the transformations are more numerous and subtle. In the main, however, there is a close relation between the amount of food consumed and the work that a man can do.

The most powerful instrument for studying the transformation of energy is still the steam engine. By means of this great invention of Watt we obtain our electrical currents; and when it is said that electricity will some day supersede steam, we can assert with a similar show of reason that flying may some day take the place of locomotion on the common road. At present there is no way of accomplishing the wished-for result. Steam produces electricity, and it is the most economical agent for producing it.

It is interesting to reflect that steam is produced by the combustion of a past vegetation. The great tree ferns of the carboniferous age required for their complex life the transformation of the sun's energy into chlorophyl and the tissues of these strange growths. This varied transformation is again made manifest in the wonderful action of coal-tar dyes in modern photography. Thus the original action of electro-magnetic waves shown in complex growths of vegetation, buried in the earth for ages, becomes evident again in a grand series of transformations. The burning of a fossilized tree trunk produces steam, steam is converted into motion, motion into electricity, and electricity into heat and light.

I have said that steam is at present our cheapest source of electricity. With the best engines it requires little more than a pound of coal to produce a horse power. This is a cheaper horse power than we can produce by water power, for the regulation and control of the supply of water is more difficult than that of a supply of steam. We thus see in New England the manufacturing towns of New Bedford and Fall River, where there is no water power, competing in number of spindles with Lowell and Lawrence, on the banks of the Merrimac River.

In studying the steps by means of which Faraday and Joseph Henry detected the small indications of a force which the steam engine has exalted into a mighty one, we are reminded of the slight rub which Aladdin gave to the lamp which was sufficient to summon a genius who could perform any tasks, from the most delicate one to the most tremendous. The feeble manifestation discovered by Faraday and Henry of what can be made a great force is called the force of magnetic induction. It may be said that it laid perdu with the possibility of being called into action by any one who possessed a coil of wire and a magnet. The reason that it remained undiscovered so long was due to the difficulty of obtaining well-insulated copper wire, and to the want of a sufficiently sensitive instrument to detect it.

The story of the discovery of magnetic induction by Faraday and Henry is most instructive, for it shows how an apparently slight and unimportant manifestation of energy can be exalted by proper means into a tremendous one. Faraday remarked, after detailing his experiments on magnetic induction: "The various experiments of this section prove, I think, most completely the production of electricity from ordinary magnetism. *That its intensity should be very feeble and quantity small can not be considered wonderful, when it is remembered that, like thermoelectricity, it is evolved entirely within the substance of metals retaining all their conducting power.*"

The steam engine has exalted this apparently feeble effect discovered by Faraday into a power which is only limited by that of the steam engine or the water power which we employ.

CHAPTER II.

MEASUREMENTS IN ELECTRICITY.

WE have said that the subject of electricity has made its great strides during the past fifty years by the intelligent application of the doctrine of the conservation of energy to it, and by the measurements of the heat equivalents of various forms of motion. In a popular treatise it is difficult to present the more or less dry details of exact measurements. I wish here merely to emphasize the fact that the mysterious force of gravitation—perhaps more mysterious in its manifestation than any electrical or magnetic force—is used to measure all our electrical manifestations.

It has been suggested, it is true, that the time of vibration of a hydrogen molecule at a definite temperature and pressure would be an unalterable standard of time, and that the wave length in vacuum of sodium light would form a standard of length independent of any change in the force of gravitation and in the dimensions of the earth. Maxwell remarks in regard to this last suggestion that "it should be adopted by those who expect their writings to be more permanent than that body."

We have what is called an absolute system of electrical measurements—absolute in the sense that everything is referred to acceleration of a body falling to the earth at a definite place through a certain space in a definite time. Absolute also in the sense that we do not deal with merely ratios.

We measure the efficiency of a dynamo which is producing the currents of electricity which are used in lighting our houses and propelling our cars by means of the mechanical equivalent of heat, which states that the work done in raising one pound of water one degree in temperature on the Fahrenheit scale is equivalent to raising 772 pounds one foot high against the attraction of gravitation. Our ultimate appeal, therefore, on the subject of the transformation of energy upon which we are entering is to gravitation, and it will be interesting at the opening stage of our study of electro-magnetism to consider gravitation as our measurer of electrical energy.

In general terms, we measure the quantity of electricity which is delivered along a wire by the current which is flowing multiplied by the time during which it flows. Now, the time is measured by a pendulum which depends for its action upon the force of gravitation. Our standard for the measure of time depends upon the pendulum, and this in turn upon the time of rotation of the earth. It is true that we may depend upon a tuning fork for the estimation of very short intervals of time, but the fork in turn is standardized by a second pendulum.

By means of measurements based upon the law of gravitation scientific men use constantly the only known universal language—that of absolute measurements. When an English-speaking physicist expresses the results of his measurements in centimetres, in grammes, and in seconds, he knows that he will be understood by a German, a Russian, a Frenchman, or an Italian; and it can be said that no other realm of human endeavour has such a universal language. It is curious to note the disposition of the human mind to dwell upon the mysteries of electricity and magnetism, and to totally ignore the greater mystery of gravitation. We are beginning to have an inkling of the relations of electricity and magnetism to light and heat and to motion. Every day fresh evidences of the laws of the transformations of energy increases our knowledge upon electricity, but we are absolutely ignorant of the relationship of gravitation to the subject of electricity and magnetism, light, heat, and motion. Gravitating force, by means of which we measure electricity, is perhaps the greatest mystery in the subject of physical science, and its manifestation is so omnipresent, so silent and unsensational, that our mind rarely dwells upon its mysterious action.

The work we have to do to overcome the force of gravitation is our measure of it. When large masses are lifted against this force we become sensible of its potency. Yet the force of attraction between two small bodies, such as two cannon balls, is extremely difficult to detect and to measure. The direct determination of the attraction between two masses by means of the common balance is the simplest way of obtaining a realizing sense of the magnitude of this force. Prof. Poynting, by many refinements, has made it also one of accuracy. The method he at first adopted was to suspend a mass from one arm of a balance by a long wire and counterpoise it in the other pan; then by bringing under it a known mass, its weight would be slightly increased by the attraction of this mass. Prof. Poynting showed that this increase in weight would be the quantity sought if the attracting mass had no appreciable effect before its introduction beneath the hanging mass, and if, when beneath it, the effect on the balance could be neglected. It was found that a mass of 453 grammes of lead hung on one arm of a chemical balance by a wire, and attracted by a mass of 154 kilogrammes of lead, showed an apparent increase of about 0·01 milligramme. We perceive from this how small the force is which is to be measured, and in order to determine it with accuracy Prof. Poynting adopted a differential method, which consisted in suspending an attracting mass from each arm of a balance instead of from one arm, and bringing another attracting mass first under one suspended mass and then under the other. By this differential method certain errors were eliminated. It was found that there was a tilting of the floor of the room in which the balance was placed, which had to be allowed for in the discussion of the results. One of the details of Prof. Poynting's investigation illustrates a refinement of modern science due to Lord Kelvin. Since the attracting force is so small, it is evident that the movements of the long pointer of the ordinary balance which indicates the difference in weight in the balance pans would be too small to observe. To make these small movements perceptible and also measurable, a small arm or bracket, B, was fixed to the pointer of the balance. One end of a spider thread or quartz fibre by which the mirror was suspended was attached to this bracket, and the other end to a fixed support, S, independent of the balance. The angle through which the mirror, M, turns for a given motion of the

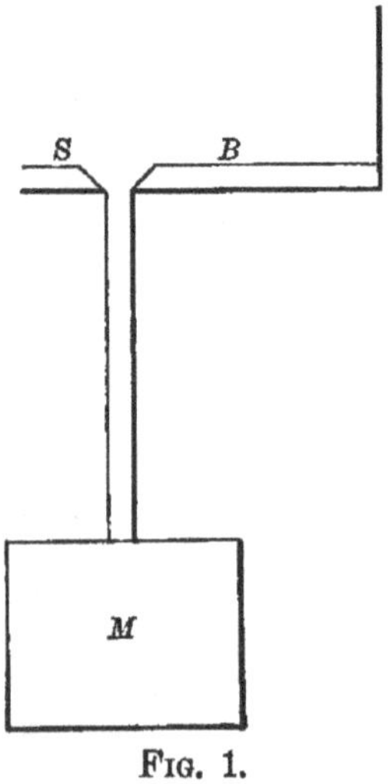

Fig. 1.

pointer is inversely as the distance between it and the fixed point, "so that by diminishing this distance the sensibility of the arrangement may be almost indefinitely increased." In Prof. Poynting's experiments, taking 4 millimetres as the distance between the threads and supposing the bracket to be 600 millimetres below the knife edge of the balance, the mirror turns through an angle 150 times as great as that through which the beam turns. The observation of the angular movement of the mirror was observed by a telescope placed at a distance. The movement of the mirror, it is evident, sweeps a beam of light through space. A movement of $\frac{1}{600000}$ of an inch could thus be detected in the motion of the attracting masses. An idea of the amount of the attraction between small bodies can also be gained from a recent investigation of Prof. C. V. Boys, who finds that the force with which two spheres weighing a gramme each (about $\frac{1}{500}$ of a pound) with their centres 1 centimetre (about $\frac{4}{10}$ of an inch) apart attract one another is nearly $\frac{7}{100000000}$ of a dyne,* and that the mean density of the earth is 5·5270 times that of water.

Although the force of attraction between bodies of small magnitude requires for its detection apparatus of extreme delicacy, yet when one of the attracting bodies is large, like the earth, the force of attraction between it and even minute bodies becomes appreciable to our senses. Take, for instance, the effect of gravitation in regulating, so to speak, the transformations of energy in our atmosphere. When water is heated, the warm water, being less dense and therefore having less

* A dyne is the force which, acting upon a gramme for a second, generates a velocity of a centimetre per second. The force of gravity acting upon a gramme generates a velocity of 981 centimetres per second.

weight, ascends in the containing vessel, while the colder water, being denser, is more strongly attracted by gravitation and descends. In this way are formed what are called convection currents in water or in our atmosphere. The force of gravitation tends to equalize the movements produced by heat. In the case of an ascending column of heated air in the atmosphere, it is evident that this air must do work in lifting the colder superincumbent air. It must therefore lose a part of its energy, or, in other words, it must be cooled in opposing the effect of gravitation.

The following interesting calculation was given in a late bulletin of the United States Weather Bureau: *

"We get 1,407 cubic miles as the average annual total of water which descends as rain or snow in the United States.

"To get some conception of this enormous mass of water we may compare it with the contents of the Great Lakes, and an approximate comparison is near enough. Lake Ontario is about 200 miles long and 70 broad, and its average depth is about 40 fathoms. It therefore contains about 636 cubic miles of water. The annual rainfall would fill it two times and leave something over for a third time. Lake Michigan is about 310 by 70 miles and has an average depth of about 50 fathoms, and consequently contains about 1,233 cubic miles of water. The average annual rainfall would fill Lake Michigan and leave 174 cubic miles over. Four years of rainfall would probably be enough to fill all the Great Lakes.

"The amount of mechanical work which the raising of this involves is enormous, and the ordinary concep-

---

* Bulletin C, 1894.

tion of it is quite inadequate. Some idea of it can be reached as follows: One inch of rain per acre makes 22,624 gallons, which equals 226,613 pounds. On a square mile the inch of water would weigh 72,516·4 tons (of 2,000 pounds each). A cubic mile of water would be this weight $\times$ 5,280 $\times$ 12 $=$ 4,593,639,104 tons, or, if the average temperature is a little above 39° F., $=$ 4,500,000,000 tons. The total weight of our rainfall (excluding Alaska) would be this multiplied by 14·07. This gives the enormous quantity of 6,332,000,000,000 tons. Let us take as a unit of handy measurement the weight of one of the lakes, say Ontario—636 cubic miles. A cubic mile of water weighs 4,500,000,000 tons. Hence, Ontario weighs 2,862,000,000,000 tons. Our average rainfall, weighing 6,332,000,000,000 tons, is therefore 2·2 Ontarios. The rain descends from clouds which average half a mile in height, and in raising the water to this height before falling Nature must perform the work of lifting 3,166,000,000,000 tons one mile per year, or 1·1 Ontarios. This, in work per day, is nearly 9,000,000,000 tons lifted one mile, and reduces to something like a lift of 100,000 tons per second. A ton lifted one mile per second is 19,200 horse power. The work done by Nature, therefore, in raising the rainfall to the clouds is equivalent to 100,000 $\times$ 19,200 horse power, or 1,920,000,000 continuous horse power, or the work of 5,000,000,000 horses working ten hours a day—perhaps a thousand times as many horses as there are in the United States.

The effect of gravitation in protecting the flora of our forests from evaporation is also very marked, for it tends to counteract the upward springing of the branches toward the light and to bend them over the earth. On certain portions of the sandy lands of Cape

Cod it is said that wooded tracts have been destroyed by cutting off the low-spreading branches of the trees and clearing up the space beneath the trees. Gravitation had done its part toward preventing evaporation. A friend, in commenting upon the very erect attitude of one who was proud of a scientific achievement, remarked that in time the force of gravitation would correct all that.

The force of gravitation gives us our units for measuring the manifestations of electricity. Does it also aid us in comprehending the forces of attraction which we perceive to be acting between magnets, and electrified pith balls?—forces which are analogous in their mathematical expression to the force of attraction between two masses, or, in other words, to the force of gravitation.

Certain students of the motion of fluids endeavor to explain gravitation by the movement of the ether. They suppose that the ether is moving throughout space—a great ether ocean with a definite tide. According to this theory, the ether passes through the sun and the planets with more or less difficulty, and its motion forces the particles of matter together.

The theory that gravitation can be explained by the motions of the ether is interesting from the point of view that it apparently unifies our conceptions of transformations of energy; for we shall see that the ether is supposed with much reason to be the medium by means of which the waves of light, heat, and electricity are conveyed to us from the sun. If we could also show that the force of gravitation results from the different rates of flow of this medium through the particles or around the atoms of bodies, we might bring gravitation into closer connection with electro-magnetism. The attempt to show that the differential flow of a medium like the ether through empty space and through space filled with material particles is one of great difficulty from both the experimental and mathematical point of view. We have never detected any effect of the ether upon the motion of bodies, and experiments upon the attraction of masses suspended in a moving fluid like water are inconclusive, since water is very different from the fluid we call ether. If we suspend, for instance, in a trough of moving water two cylinders of gauze of different mesh, these cylinders will be apparently attracted to each other by degrees varying with the difference in the fineness of the meshes of the gauze. We must reflect, however, that we can not reason from the motions of a fluid possessing viscosity, like water, to the motions of an ether which is not viscous.

If we suppose that the attraction of gravitation arises from a stress in the ether, this stress would have to be enormous.

Williamson * calculates that the amount of the ether stress at the earth's surface to account for gravitation would be 4,000 tons on the square inch.

"Lord Kelvin has shown that if we suppose all space to be filled with a uniform incompressible fluid, and if we suppose either that material bodies are always generating and emitting this fluid at a constant rate, the fluid flowing off to infinity, or that material bodies are always absorbing and annihilating the fluid, the deficiency flowing in from infinite space, then in either of these cases there would be an attraction be-

* Introduction to the mathematical theory of the stress and strain of elastic solids.

tween any two bodies inversely as the square of the distance. If, however, one of the bodies were a generator of the fluid and the other an absorber of it, the bodies would repel each other." Maxwell,* in criticising this supposition, remarks that it seems to require an absorption or annihilation of matter.

An attempt to explain the force of gravitation by the impact of corpuscles was brought forward by Le Sage, who supposed that all bodies are bombarded by an immense number of corpuscles which are flying about with great velocity. One mass, therefore, partially shades or protects another neighbouring mass from the impacts of such corpuscles, and the masses being more bombarded on the sides that are not opposed are apparently attracted toward each other. This theory gives a very plausible explanation of the mysterious force of gravitation; but Clerk Maxwell, applying the doctrine of the conservation and transformation of energy, proved that the theory must be supplemented by other theories which are as little supported as the theory itself.

Maxwell shows that the velocity of Le Sage's corpuscles must be enormously greater than that of any of the heavenly bodies, otherwise they would act as a resisting medium and oppose the motion of the plants. The energy of the corpuscles would be enormous. The rate at which it must be spent in order to maintain the gravitating property of a single pound is at least millions of millions of foot pounds per second. If any appreciable amount of this energy is communicated to a body in the form of heat, the amount of heat so generated would raise the whole material universe in a few seconds to a white heat.

---

* Encyclopædia Britannica, Gravitation.

"Prof. Challis has investigated the mathematical theory of the effect of waves of condensation and rarefaction" in an elastic fluid on bodies immersed in the fluid. He concludes that the effect of such waves would be to attract the body toward the centre of agitation or to repel it from that centre, according as the wave's length is very large or very small compared with the dimensions of the body.

A tuning fork set in vibration attracts a delicately suspended body. Lord Kelvin shows that in fluid motion the average pressure is least when the average energy of motion is greatest. The wave motion is greatest near the fork and the pressure there is least, and the suspended body is therefore urged toward the fork. Maxwell remarks that all the theories which have been brought forward to explain gravitation—namely, Le Sage's corpuscle theory, the generation or absorption of fluid by bodies under pressure, the wave theory—require the expenditure of work. "According to such hypotheses we must regard the processes of Nature not as illustrations of the great principle of the conservation of energy, but as instances in which, by a nice adjustment of powerful agencies not subject to this principle, an apparent conservation of energy is maintained. Hence we are forced to conclude that the explanation of the cause of gravitation is not to be found in these three hypotheses."

It is interesting to follow the working of Faraday's mind on the subject of gravitation. Having opened a great field in the transformations of electricity into magnetism, and of magnetism into electricity, his mind sought to embrace the force of gravitation in a generalization which should include it with those of electricity. An experiment by Faraday is always worthy of respectful consideration; but in following him in this excursion we feel in a certain sense like one who accompanies a voyager on a strange sea. There was nothing to guide him. No experiments, no shadowy intimations of relationship, such as had always accompanied the manifestations of electricity and magnetism. His mind naturally rested upon the electrical methods which had proved so fruitful in discovering the laws of induction, and he arranged his apparatus as follows. Since he knew nothing about gravitation except its measure and the direction in which it acted, he determined to apply the analogies of the laws of the transformation of energy, to see if some electrical work was not done when a ball of glass or wood was allowed to fall along the lines of gravitation and was immediately drawn up against the force of gravitation. It is evident that his fruitful conception of space filled with lines of magnetic and electric force occurred to him. A coil of wire moved in a certain manner across magnetic lines of force showed electrical disturbances. Such a coil, however, did not give any indications, no matter how it was moved in reference to the lines of force, of gravitation. The needle of a delicate galvanometer remained absolutely quiet if the galvanometer were connected with a coil which was moved across or along the lines of gravitating force. What led Faraday to suppose that an effect could be observed by allowing a nonmagnetic body like glass to fall through the coil and then to be drawn up through the coil against gravitation? In one case work was being done by the falling body, and in the other case work was done against the pull of gravitation. It is evident that Faraday expected the electrical equilibrium of the coil to be disturbed by this transformation in the space outside it.

Faraday, in speaking of his efforts, remarks:

"The long and constant persuasion that all forces of Nature are mutually dependent, having one common origin, or rather being different manifestations of one fundamental power, has made me often think upon the possibility of establishing by experiment a connection between gravity and electricity, and so by introducing the former into the group, the chain of which, including also magnetism, chemical force, and heat, to bind so many and such varied exhibitions of force together by common relations."

"Hypothesis: Two bodies moved toward each other by the force of gravity currents of electricity might be developed either in them or in the surrounding matter in one direction; and that, as they were by extra force moved from each other against the power of gravitation, the opposite currents might be produced. A body was allowed to fall with a helix, and afterward through a helix. Various bodies were used, notably copper, bismuth, glass, sulphur, gutta-percha. It was thought that the stopping of the up-and-down motion in the line of gravity would produce contrary effects to the coming on of the motion, and that whether the stopping was sudden or gradual; also that a motion downward quicker than that which gravity could communicate would give more effect than the gravity result by itself, and that a corresponding increase in the velocity upward would be proportionally effectual. A machine was devised which could give a rapidly alternating up-and-down motion.

"Here end my trials for the present. The results are negative. They do not shake my strong feelings of the existence of a relation between gravity and electricity, though they give no proof that such a relation exists."

We shall now enter upon the study of the mysteries of electricity by means of the more mysterious force of gravitation; for, as I have said, our measurements of force in general are based upon measurements of the force of gravitation. Our great unit in the study of transformations of energy is the mechanical equivalent of heat, and, as we have already pointed out, this is expressed in terms of a mass lifted a certain height against the force of gravitation.

CHAPTER III.

MAGNETISM.

To an American, the question What is electricity? has a great national interest, for Count Rumford, Benjamin Franklin, and Joseph Henry have been like electric lights spaced on a dark path of a high mountain, and we are slowly ascending to a summit, having been guided by the rays of their genius. In America, too, the practical applications of electricity have extended with such swiftness that one might say that there is something in the manifold transformations of electricity peculiarly congenial to the American temperament. While the reader, however, may be willing to admit the claims of Benjamin Franklin and of Joseph Henry to be pioneers in the subject of electricity, he may doubt the justice of including Count Rumford with these workers on the question, What is electricity?

Our final answer to this great question will necessarily embrace the labours of Count Rumford in the subject of heat; for his celebrated experiment on the heat developed in the boring of a cannon in Munich drew mens' attention to the transformation of mechanical work into heat, and led them to reflect on the conservation of energy. Guided by this great theory, we have learned that light, heat, electricity, and magnetism can be studied and embraced under one head—that of the transformations of energy. Electricity no longer stands apart, a mysterious force, as Franklin regarded it, having no connection with light or heat. It is now seen that we can not study it apart from the manifestations of the latter.

Knowing that Count Rumford, when a boy, walked from Woburn to Cambridge, a distance of eight miles, to attend the lectures of Prof. John Winthrop, the first Professor of Physics at Harvard University, I was interested to ascertain how much he learned of the subject of electricity. In the college archives I found a time-worn notebook in the handwriting of Winthrop, and among the notes of excellent lectures on astronomy and a few on light and heat I found, apparently, that but one lecture had been given on magnetism and one on electricity. As a curious illustration of the extent of our knowledge of a great subject less than a century and a half ago, I give the main portion of his notes on the lecture on electricity in 1750: "If a flaxen string be extended and supported, and at one end an excited tube be applied, light bodies will be attracted, and that at the distance of 1,200 feet at the other end. This electricity since the year 1743 has made a considerable noise in the world, upon which it is supposed several of the (at present) hidden phenomena of Nature depend. . . . Men have been so electrized as to have considerable light round their heads and bodies, not unlike the light represented around the heads of saints by the painters."

The entire apparatus to illustrate the subject of electricity and magnetism in Harvard University until the year 1820 consisted merely of two Franklin electrical machines, a collection of Leyden jars, and small apparatus to illustrate the effects of electrical attractions and repulsions shown by electrified pith balls or similar light objects. I have had the Franklin machine photographed beside a modern electrical machine which can be carried around in the arms, and which has many times the efficiency of the machine employed by Franklin. Fig. 2 shows a cut from this photograph. During the days, therefore, of Prof. Winthrop the knowledge of electrical phenomena was extremely small. It was confined to the observation of the attrac-

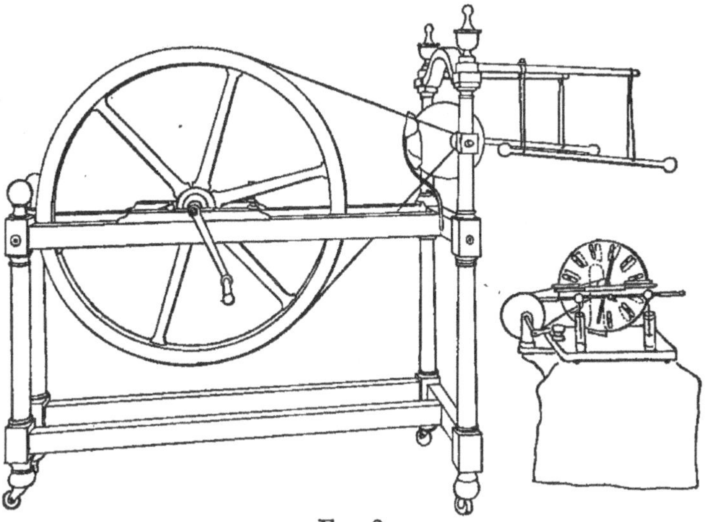

FIG. 2.

tion of magnets and of the phenomena of frictional electricity. America, however, in the year 1750 knew as much as Europe, and the physical cabinet of Harvard University was not more poverty-stricken than that of the University of Leyden. Our advance since 1750 has been due to the accurate quantitative investigation of the transformations of energy; and although Benjamin Franklin's brilliant experiment in establishing the identity between the manifestations of lightning and those of the ordinary electrical machine is often referred to as the beginning of our real knowledge of electricity, I should say that the experiment of Count Rumford in boring the cannon has had far more real influence in the development of true ideas in regard to the transformations of energy in which electricity plays such an important part.

In hitherto unpublished letters of Count Rumford to Pictet, of Geneva, in 1797, now in the possession of the American Academy of Arts and Sciences, Boston, he shows how clearly he had seized upon a fundamental idea of the transformations of energy:

"Your friend Mr. Joly will perhaps have mentioned to you a late experiment of mine in which I caused more than four gallons of water to boil without fire—merely by the heat generated by the friction of two metals rubbed against each other. I have just finished a calculation by which it appears that the heat generated equably, or the stream of heat which flowed with a uniform velocity—if I may so express myself—was, in one of my experiments, equal to that generated equably in the combustion of nine middling-sized wax candles all burning together or at the same time.

"As the machinery which produced the friction which generated this heat could easily be put and kept in motion by the strength of one strong horse, we see how much heat could be generated by the strength of animals, without either fuel, light, or chemical decomposition.

"I am just now engaged in writing a paper on the subject which I mean to send to the Royal Society. The results of my experiments seem to me to prove to a demonstration that there is no such thing as an igneous fluid, and consequently that caloric has no real existence. You must not, however, raise your expectations too high respecting my experiments. Though they were made on a large scale and conducted with care, there was nothing very new or very remarkable about them; and as to their results, they prove only this single fact (of which most probably you never had any doubt), that the heat generated by friction is *inexhaustible*, even when the bodies rubbed together are to all appearance perfectly insulated or put into a situation in which it is evidently impossible for them to receive from the other bodies the heat they are continually giving off.

"It appears to me that *that* which any insulated body or system of bodies can continue to give off without limitation can not be a material substance. A bell when struck with a hammer gives off sound, but I do not think it would be speaking philosophically to call sound a material substance."

In another letter to Prof. Pictet, dated Paris, May 4, 1804, he says, "I am persuaded that I shall live a sufficiently long time to have the satisfaction of seeing caloric interred with phlogiston in the same tomb."

The science of electricity took an immense stride as soon as the transformations of energy were studied from a mechanical standpoint—in other words, from the standpoint of Count Rumford—and as soon as men abandoned theories of subtle fluids and began to measure the forces of attraction and repulsion and the equivalence between motion and the energy it makes manifest, whether we convert this energy into heat or into electricity. Count Rumford saw clearly only the transformation of mechanical work into heat, and the relation between the work of a horse in producing this transformation and the food which he eats. Indeed, he made a rough calculation of this transformation. We shall see that as soon as Faraday showed that motion could also be converted into electricity, and when Joule showed the equivalence between the energy of movement and the electrical energy produced, we entered upon the new era of electricity—an era which is characterized by our study of the transformations of energy.

In the subject of electricity delicate measuring instruments have played a most important part. In general these instruments measure attractions and repulsions. Before the year 1800 there were no delicate instruments for measuring such forces. A compass on a pivot was the most sensitive instrument that was used to study magnetism, and the electrified pith balls or the suspended gold leaves constituted the measuring apparatus in all that was known then of electricity. The subject of electricity took its great stride not from the use of the instruments employed by Franklin or from the side of the subject investigated by him, but rather from the side of magnetism. It was the movements of a magnetized needle that led Faraday to his great discovery of induction and the conversion of motion into electricity. We have learned since 1830 a great deal about the magnetic properties of soft iron under electrical influences, we know, however, little more of the properties of the loadstone and of permanent magnets, and of the magnetism of the earth, than was known to Count Rumford or to Sir Isaac Newton. Both of these philosophers, I imagine, thought that if all the loadstones in the world, the earth's magnetism, and the permanent magnets, were destroyed that magnetism would disappear from the sum of the forces whose origin and manifestations perplex the human mind. They had no conception of the possibility of producing a magnetic condition in iron by means of a wire wrapped around the iron and connected with a voltaic cell; for the principles of electro-magnetism were not discovered until 1819. To-day, if there should be such a destruction as we have outlined—namely, that of loadstones, permanent magnets, and the disappearance of the earth's magnetic force—the world would be apparently only incommoded by the substitution of observations on the sun and stars for the observation of the ship's compass. We could produce permanent magnets and powerful electro-magnets at pleasure and in any quantity by the use of voltaic cells or the dynamo.

The question, however, What is magnetism? is closely allied to the question, What is electricity? and before entering upon the phenomena of electro-magnetism it is well to consider the force of magnetism.

In a letter to M. Dubourg, dated London, March 10, 1773, Franklin thus explains his views of magnetism:

"SIR: As to the magnetism which seems produced by electricity, my real opinion is that these two powers of Nature have no affinity with each other, and that the apparent production of magnetism is purely accidental. The matter may be explained thus:

"1. The earth is a great magnet.

"2. There is a subtle fluid, called the magnetic fluid, which exists in all ferruginous bodies, equally attracted by all their parts, and equally diffused through their whole substance; at least where the equilibrium is not disturbed by a power superior to the attraction of the iron.

"3. This natural quantity of the magnetic fluid which is contained in a given piece of iron may be put in motion so as to be more rarefied in one part and more condensed in another; but it can not be withdrawn by any force that we are yet made acquainted with so as to leave the whole in a negative state, at least relatively to its natural quantity; neither can it be introduced so as to put the iron into a positive state or render it *plus*. In this respect, therefore, magnetism differs from electricity.

"4. A piece of soft iron allows the magnetic fluid which it contains to be put in motion by a moderate force, so that, being placed in a line with the magnetic pole of the earth, it immediately acquires the property of a magnet, its magnetic fluid being drawn or forced from one extremity to the other; and this effect continues as long as it remains in the same position, one of its extremities becoming positively magnetized and the other negatively. This temporary magnetism ceases as soon as the iron is turned east and west, the fluid immediately diffusing itself equally through the whole iron, as in its natural state."

In the same letter he still further enforces his fluid theory of magnetism.

The molecular theory of magnetism, which may be said to appeal strongly to chemists, since it seems more or less in consonance with the theory of attraction of molecules, has been advanced by various physicists, particularly by Weber. This theory supposes that the molecules of a bar of iron are small magnets which, when the bar is unmagnetized, point indiscriminately in all directions, but when it is magnetized a certain number of these little magnets point in a definite direction. Many of the phenomena of magnetism give great colour to this theory, and it may be said to be satisfactory until we examine the action of magnetism on light; then we perceive that there are not only attractive forces between molecules of the iron, but also rotary motions in the medium within and around the magnet, and in order to comprehend the phenomenon of magnetism we are compelled to assume a medium between the molecules and to attribute rotary or vortex movements to it. We shall return to this more comprehensive theory later. The truth seems to lie between the two theories. It is certain that the molecular condition of iron and steel greatly influences its magnetic condition. For instance, if a permanent steel magnet is struck with a hammer it loses a part of its magnetism. The molecules appear to be shaken out of the position which they had taken on being magnetized. If a bar of soft iron is held in the line of magnetic dip—that is, the line indicated by a long magnetic needle suspended from its middle, the line along which a south pole, if it could be separated from its north pole, would travel toward the north pole of the earth—and is struck a sharp blow with a hammer, it will become magnetic. Its condition can be tested by presenting its ends to an ordinary compass. If, on the other hand, it is held east and west and struck, it can be deprived of its magnetic state. It is found that iron ships become magnetic in certain portions under the hammering and vibrations to which they are subjected in shipyards. If the molecules of steel are subjected to high temperature—for instance, to a white heat—the steel is deprived of its magnetism. Very interesting work in this field has been done by Profs. Hopkinson and Ewing. Some years since I examined the behaviour of a piece of steel when cooled to 80° C. below zero. I found that its magnetic condition, measured by what is called its magnetic moment, was diminished 50 per cent. In calling attention to this observation I remarked that every piece of steel had its idiosyncrasies, and I hoped to continue my observations with other specimens. Prof. Dewar, who has made some remarkable observations on molecular conditions at very low temperatures, has lately examined the effect of great cold on magnetism, and I can not do better than to quote his words in a paper on the scientific uses of liquid air:

"Prof. Trowbridge examined the effect of a temperature of −80° C. upon a permanent magnet, and came to the conclusion that the magnetic moment was diminished by about 50 per cent. Prof. Ewing found that an increase of temperature of 150° C. above 10° caused a reduction of the magnetic moments of a bar magnet by about 40 per cent, and that the magnet on cooling recovered its original state. This result would lead us to expect that if the same law is followed below the melting point of ice as Ewing found above it, then a bar magnet cooled to −182° C. ought to gain in magnetic moment something like 30 to 50 per cent. The experiment of Prof. Trowbridge is, however, apparently opposed to such an inference. It appears that Prof. Trowbridge cooled a magnet that had not reached a constant state (that is to say, one that on cooling would not have completely recovered its magnetization on cooling), because, after the magnet had been cooled to −80°, on regaining the ordinary temperature it had lost 50 per cent of its original magnetic moment. Such a magnet would apparently diminish in magnetic moment on cooling and heating the first time the action was examined, but a repetition of the process when the action of magnetization and temperature were strictly reversible might lead to an opposite conclusion. To settle this question, a series of experiments on the magnetic moment of small magnets cooled to −182° were carried out. Small magnets from half an inch to an inch in length were made of watch spring or steel wire, and were either used separately or in bundles. . . . After the first cooling the magnet is allowed to regain the ordinary temperature, and the operation of cooling and heating is repeated three or four times."

In summing up his results Prof. Dewar finds that every magnet has individual characteristics that may either result in no change on cooling or the addition or subtraction of from 12 to 24 per cent in the magnetic strength. His work bears directly upon the molecular theory of magnetism. Prof. Ewing has also shown how magnetic models consisting of an aggregation of magnets suitably suspended can represent most of the phenomena of magnetism, irrespective of rotary phenomena.

Before the year 1800 the compass represented to the world the principal use of magnetism; later magnetism became like the genius which Aladdin called into existence by motion. Furthermore, the force of attraction between the poles of two magnets was considered as an action at a distance, without the intervention of any medium between the poles. We shall see that the great advances in our knowledge of electricity and magnetism have come from our study of the actions in the medium between two magnetic poles or two electrified pith balls.

The general phenomena of magnetism apart from electro-magnetism were known at an early period, and the earth was recognised as a great magnet. It was observed that all pieces of iron and steel became more or less magnetic if they were left standing at an angle with the horizontal plane, which in this latitude is about 73°. A poker, for instance, suspended by a fireplace will generally show two poles. The singular fact that there is a limit to the length of a magnet was early commented upon. It is impossible to maintain the poles of a magnet ten feet from each other on a steel bar; intermediate poles will generally form, which are called consequent poles. My attention, a few years since, was called to a magnetic motor which well illustrates the formation of consequent poles. An inventor claimed that he had discovered a neutral line in the space near the poles of a horseshoe magnet. His motor consisted of an armature of soft iron (N S, Fig. 3) on the end of a balance arm. The other end of the arm vibrated between two stops. It was claimed that when the proper rate of vibration was reached the armature would move across the neutral line to and fro, the neutral line acting as a species of cut-off. The existence of the neutral line was shown by presenting small tacks to a bar of iron which was moved in front of the poles of the magnet. As the bar was moved away from the poles the tacks dropped off; but in continuing the motion away from the magnet another position was found in which the tacks again were attracted to the bar. The bar apparently had moved through a neutral line. The phenomenon, however, was caused by the shifting of the consequent poles, due to the changing intensity of the

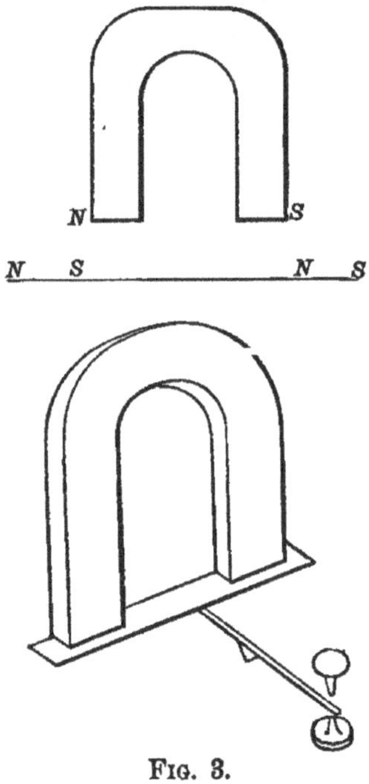

FIG. 3.

field, and had no real existence. It is needless to say that the motor did not run from this cause.

For many years there was no substantial change in the form of the ship's compass. One or more comparatively heavy bar magnets were affixed to a card bearing the points of the compass, and the card was pivoted so that its graduations should pass fixed points on the compass box. Lord Kelvin made a great improvement in the old form by fixing a number of very light steel magnets on a light disk, getting in this way strong magnetism, so to speak, or magnetic moment combined with extreme lightness and steadiness. The employment of iron ships has made it necessary to compensate the attraction of the vessel upon the compasses by placing steel magnets in proper positions on the deck near the compass, or by placing a compass high above the deck, and by means of its indications correcting the lower compasses. It is necessary also to swing the ship occasionally—that is, to take the bearings of some point on the shore near a harbour while the ship is turned completely around. In this way the attraction of the mass of the ship on the compass when sailing on different courses is ascertained.

The force between two attracting pith balls can be represented by $F=\frac{m\ m_1}{r^2}$ where m and m_1 are the charges on the pith balls and r is the distance between them. When there is a plate of glass between the charged balls the force is very much diminished. If k is a factor depending on the insulating power of the glass, the force is $F=\frac{m\ m_1}{k\ r^2}$; that is, the greater the insulating power the smaller the force. In a similar manner we can express the force of attraction between two magnetic poles as $F = \frac{m\,m_1}{r^2}$, in which m and m_1 represent the strength of the poles and r is the distance between them. If, however, the poles are immersed in a solution of iron the attraction between them would be expressed by $F = \frac{m\,m_1}{\mu\,r^2}$, in which μ is a quantity which depends upon the medium. If μ is large, the force of attraction is small. The attraction, for instance, between two magnets placed in a vessel containing a salt of iron would be less than in air.

It is to such considerations of the nature of the surrounding medium that we owe the advances in our knowledge of magnetism. Previous to the year 1800, I have said, no account was taken of the surrounding media, and one magnetic pole was considered to act upon another as if it were an action at a distance and not from point to point in the medium between the attracting poles. It is well to consider these two points of view carefully, for in the present study of electricity and magnetism we are chiefly occupied with a study of what goes on in the medium in which the attracting bodies are immersed. The early workers in magnetism apparently formed no mental picture of the state of the field in which the attracting bodies were placed. They had not formed the conception of lines of force, although they were familiar with the arrangement of iron filings around the poles of a magnet. The defenders of the theory of action at a distance did not carefully examine the disturbance of a magnetic field by the introduction of a magnetic pole, whereas in the modern theory of action from point to point in the medium this disturbance is fully considered. All space around the earth is filled with lines of force, which crowd into the north and south pole. In the room in which the reader of this book is seated these lines are generally parallel and the space of the room is filled with them. If there is an iron article in the room they tend to crowd through the iron, finding it a better conductor than the air. We say, therefore, that there is a flow of magnetic induction through the iron. We can think of magnetic induction as the accumulation of the lines of force in the space occupied by the iron. To obtain a space anywhere on the earth's surface free from lines of force we must employ a thick-walled hollow sphere of iron. In the space inside the sphere there will be no lines of force. The space has been, so to speak, swept free of lines of force, and there is a flow of induction through the walls of iron. A compass inside such a sphere would no longer be influenced by the poles of the earth. An examination of the terminology of a subject often will show the trend of the subject. We speak to-day of the flow or flux of magnetic induction, and we picture to ourselves this flow as taking place from one pole of a magnet to the other through the air. We also speak of closed magnetic circuits, which we obtain to a great extent when we put an armature on a horseshoe magnet. The flow of induction takes place then through the armature, and not through the air. A compass is not affected by a closed magnetic circuit, for no lines of force escape to flow through its steel. We also speak of the resistance of the air to the flow of magnetic induction. It is harder to force the flow of magnetic induction through air than through iron or steel. If, for instance, we should cut a horseshoe magnet in two at its bend and then put the ends together again, it can not be made as strong as it was originally; although the joint may be made very perfect, the thin layer of air opposes a resistance to the flow of induction. We also speak of the magneto-motive force which establishes the flow of induction, much as we speak of the electro-motive force which establishes a current of electricity. In a certain sense, therefore, we can regard a magnet as a little battery. It has a magneto-motive force. It has a circuit with a certain resistance. It has a flow analogous to a current.

Although this terminology shows that we are studying the action from point to point in the substance of a magnet and in the space around it, we must not conclude that we have evidence of a flow of a magnetic fluid or the flow of the ether of space around and through the substance of a magnet. We shall see, also, that we have no evidence of an actual flow of electricity in the case of an electrical current.

It is often stated that certain persons have detected a peculiar effect when the poles of a magnet are passed over their bodies, and that they have seen lambent flames proceeding in the dark from the poles of a magnet. The reader curious in this matter will find many observations of these alleged phenomena in a book entitled The Odic Force, by Baron Reichenbach.* It has been found, however, that wooden bars painted to resemble magnets produce the same effect as the actual magnets. The effect of very powerful electro-magnets on the human body has been tried recently. With the head placed between the poles of an electro-magnet sufficiently strong to lift tons of weight one can carry on mental calculations with perfect ease, and a genius undisturbed by the *milieu* doubtless could write a sonnet to magnetism. The human nervous system does not appear to be sensitive to magnetic force.

---

* Edited by John Ashburner, M. D. Partridge & Brittan, New York, 1855.

On the other hand, all bodies may be considered more or less magnetic. The oxygen of the air is highly so compared with other gases. If the air were deprived of its oxygen magnetic poles would attract each other more strongly. The early electricians divided all bodies into two classes, paramagnetic and diamagnetic. Iron and steel, nickel and cobalt, are paramagnetic bodies. Bismuth is an example of a diamagnetic body. A bar of bismuth will place itself at right angles to the line between a north and a south pole, instead of in this line, as a bar of iron or steel would do.

In the case of electricity and magnetism the medium between conductors of electricity or between magnets plays an important part in the phenomena of what is called induction—in other words, the phenomena which result from the stresses called forth in the medium by the electro-magnetic waves. In gravitation, the medium, however, does not affect the attraction between the bodies. A plate of iron or a plate of paraffin between two attracting spheres does not affect their pull upon each other. There is also no evidence of a directive action of gravitation—that is, an elongated cylinder does not tend to place its longer axis vertical on the earth's surface.

Maxwell shows that "if we assume that the medium is in a state of stress, consisting of tension along the lines of force and pressure in all directions at right angles to the lines of force, the tension and the pressure being equal in numerical value and proportional to the square of the intensity of the field at the given point, the observed electrostatic and electro-magnetic forces will be completely determined. . . . A path is now open by which we may trace to the action of a medium all forces which, like the electric and magnetic forces, vary inversely as the square of the distance, and are attractive between bodies of different names, and repulsive between bodies of the same names."

Gravitation differs from electrical and magnetic attractions in that there is no difference of sign, no positive and negative. We must therefore assume a different kind of stress. "We must suppose that there is a pressure in the direction of the lines of force, combined with a tension in all directions at right angles to the lines of force. Such a stress would no doubt account for the observed effects of gravitation."

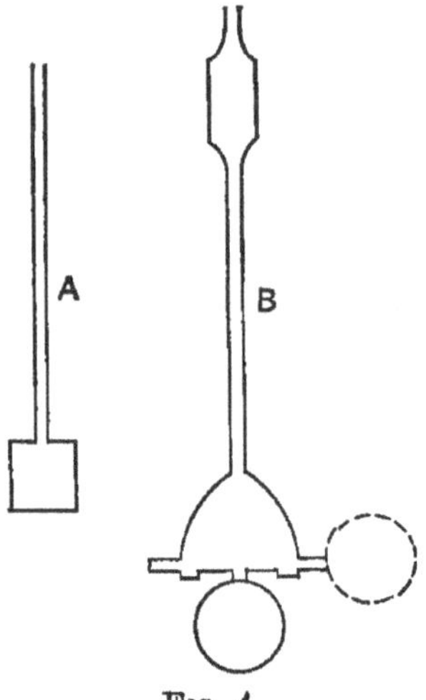

FIG. 4.

Prof. C. A. Bjerknes,* in a very suggestive paper on Hydrodynamical Analogies, has shown that magnetic and electrical attractions and repulsions can be imitated by giving fluids a pulsating or oscillating movement. By the term pulsation is meant a periodical change of volume; by that of oscillation, a change of length. Bjerknes' pulsator consists of a little tambour or drum (A, Fig. 4), covered on both ends with membrane. The drum, A, is connected by rubber tubing to an alternating air pump, which communicates air pulsations to the drum. The oscillator, B, consists of a sphere supported on a little lever; this lever can be set in vibration by the alternating air pump, communication being obtained by means of a tube. The pul-

* Repertorium der Physik, xix, 1883.

sators and the oscillators were mounted on suitable balance arms, and were immersed in fluids. Thus, on working the alternating pump pulsating or oscillatory movements could be communicated to these fluids.

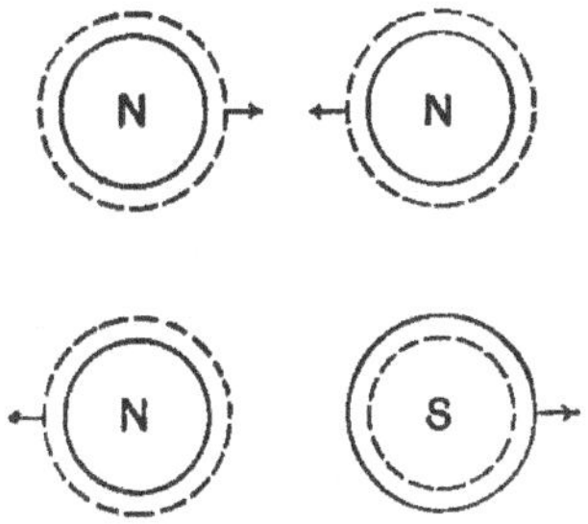

FIG. 5.

When two pulsators are brought near each other and when the pulsations are in the same phase there is an attraction (Fig. 5). On the other hand, if the pulsations are discordant—that is, in opposite phases—repulsion occurs. This is the exact opposite to what takes place in magnetism where like poles repel and unlike attract. Experiments were next tried with a combination of pulsators and oscillators. If an oscillating sphere was brought near a pulsator, attraction or repulsion could be obtained—attraction, if the oscillating sphere approached the pulsator at the instant of its dilatation, or repulsion if the sphere was removed from the pulsator in the time of its dilatation. Here we obviously have the means of drawing many interesting analogies between the motion of fluids and the phenomena of electricity and magnetism. Bjerknes shows that the phenomena of induction, temporary and permanent magnetism, and the phenomenon of the magnetic field can be imitated by his hydrodynamic apparatus.

Although analogies drawn from the action of fluids are highly interesting, they remain at the present time in the subject of electro-magnetism merely analogies. Fluids are gross substances, and their molecular grouping and attractions are of such magnitude that we can not reason safely from actions in them to actions in a highly attenuated fluid or medium, such as that which is supposed to exist in the space about electrical conductors and magnets. Wave motions in the ether are supposed to take place around the atoms or molecules of bodies. These wave motions are handicapped, so to speak, by the presence of such bodies and by the waves which the vibrations of the bodies impress on the ether. Bjerknes's experiments have led many philosophers to speculate upon the possibility that the pulsation or oscillatory movements of an ether might account for the phenomena of gravitation and of attraction or repulsion in general.

Lord Kelvin, in an address on terrestrial magnetism, remarks, "I find it unimaginable but that terrestrial magnetism is due to the greatness and rotation of the earth"; and says further in regard to the hypothesis that magnetic disturbances are caused by the sun acting as a variable magnet: "In eight hours of a not very severe magnetic storm as much work must have been done by the sun in sending magnetic waves out in all directions through space as he actually does in four months of his regular heat and light. This result, it seems to me, is absolutely conclusive against the supposition that terrestrial magnetic storms are due to magnetic action of the sun; or to any kind of dynamical action taking place within the sun, or in connection with hurricanes in his atmosphere, or anywhere near the sun outside. It seems as if we may also be forced to conclude that the supposed connection between magnetic storms and sun spots is unreal, and that the seeming agreement between the periods has been a mere coincidence." *

* Popular Lectures and Addresses, by Sir William Thomson (Lord Kelvin).

CHAPTER IV.

THE ELECTRIC CURRENT.

A DISTINGUISHED American lecturer once delivered a brilliant course of lectures on the Lost Arts, in which he showed that the ancients possessed a knowledge of many processes now lost to the world, and he intimated that the nation that could transport immense obelisks over great distances, and build the pyramids, might also have had a knowledge of science far beyond what historical investigation has revealed to us. It may seem unsafe to assert positively that the ancients knew nothing of electricity save what is manifested by the rubbing of amber; for modern excavations are continually increasing our knowledge of ancient life. Not the slightest evidence, however, exists that a knowledge of the art of covering a wire uniformly with silk, cotton, or other nonmetallic or insulating substance existed before the time of Benjamin Franklin.

Without insulated wire no progress was possible in electricity, for the electro-magnet, by means of which we transmit telegraph messages, propel cars, and, in short, accomplish most of the modern wonderful results in electricity, consists merely of a spool of cotton- or silk-covered wire placed upon an iron core. The rapid development of electrical science during the past thirty years is largely due to the facility with which a piece of iron can be wound with covered wire. Prof. Joseph Henry, whose electrical researches led to the invention of the telegraph, spent months in wrapping wire with strips of cloth, in order to make the magnets by means of which he showed the possibility of transmitting signals to a distance by electricity. To-day far more powerful magnets than he constructed can be made in an hour.

This wonderful something which we call electricity circulating around coils of covered wire makes an iron core a magnet. The oftener we make it flow around the iron, or, within certain limits, the more turns of wire we put upon our spool, keeping the strength of the electric current constant, the stronger the magnet. Now, in considering this phenomenon we are led to the remarkable fact that a thin covering of silk or cotton can prevent the electric current, such as is used to propel cars, from being conducted from one layer of wire on the spool to the next. The thinnest sheet of paper placed under the trolley of an electric car will stop the car, provided that it is not punctured by the mechanical pressure. Cut a copper wire carrying an electrical current, file the ends square, place a sheet of writing paper between the ends and press them together: the current which was transmitting thousands of horse power is stopped; it is incapable of passing through the insulating substance of the paper. Here we are brought to a realizing sense of one of the chief peculiarities of the method of transmitting power by electricity. No other agency for transmitting power can be stopped by such slight obstacles as electricity. A sheet of writing paper placed across a tube conveying compressed air would be instantly ruptured. It would take a wall of steel at least an inch thick to stand the pressure of steam which is driving a 10,000-horse-power engine. A thin layer of dirt beneath the wheels of an electric car can prevent the current which propels the car from passing to the rails, and thus back to the power house.

Another striking difference between what we call a steady electrical current in a wire and steam or air at high pressure in a pipe is the absolute stillness which marks its sudden passage from one wire to another one of larger diameter. In the case of air or steam, this sudden passage from a small conductor or pipe to a larger receptacle would be accompanied by a whistle or roar. There is no sound heard in a wire when an electric current is suddenly established in it. One can handle the wire carrying thousands of horse power without experiencing the slightest sensation, and birds can sit on such a wire with the safety that they can rest on a dead limb of a tree. When the wire is broken, however, there is a blinding flash of light and a loud report—a miniature thunderstorm. The only analogy between our pipes conveying air or steam at high pressure and our wire carrying an electrical current is in the heat which we can perceive on both the pipes and the wire. This analogy is a valuable one, which we must bear in mind as we endeavour to discover what electricity is. The transmission of compressed air or of steam by pipes is not affected by the presence or the motion of surrounding objects outside the pipes; a compass needle remains quiescent. The neighbouring pipes are not attracted or repelled from each other. This is not true in the case of the electrical transmission of power. A compass needle instantly points to the wire carrying the current; and iron dust or filings gather around the wire. Two neighbouring wires carrying currents are attracted to each other if the currents are going in the same direction, or repelled if they are going in opposite directions. The quick removal of one of these wires from its proximity to the other makes the currents in the wires throb, just as a change in the pressure of the air or steam in a pipe would cause a fluctuation in the transmission of power.

The pressure of compressed air or steam is not sensibly affected by a rapid motion of the pipe containing it; for instance, we could quickly revolve a hollow ring like a top about a diameter, allowing steam at high pressure to pass into the ring through a suitable valve at one end of the diameter and out of the ring at the other. The whirling motion of such a moving ring would not sensibly affect the air or steam in the pipe. If a current of electricity, however, should enter and leave a ring made of wire it could be very much affected by the rapid motion of the ring. The quick rolling of a steamship does not modify the pressure of the steam in the pipes conveying it to the engines; it does, however, affect the electric current which is lighting the steamship, although the effect in the case of the comparatively slow motion of the steamship is small. When a loop or ring of copper wire carrying a current is revolved many times a minute before the pole of a powerful magnet a fluctuating effect can be produced which would be sufficient to make the electric lights of the steamship fed by this current blink painfully. The pole of the earth exerts a similar effect on moving wires carrying currents.

The mysterious something which we call an electric current is therefore influenced by the motion of surrounding objects and by the motion of its own path; it has little in common with the compressed air or steam which is conveying an equal amount of horse power, except in the evidence of heat along its path.

We have said that the only peculiarity which a pipe conveying steam or compressed air possesses in common with a copper wire conveying an electric current resides in the development of heat along the conductor. If we should narrow the bore of a steam pipe to the size of a knitting needle, we should greatly restrict the flow of steam through the pipe and reduce the amount of horse power that was previously transmitted. The same is true of a water pipe: the flow of water is greatly impeded by the constriction of the pipe. In the case of steam and compressed air there is not a notable or great increase of heat when the bore of the conducting pipe is made smaller. With electricity, however, there is a remarkable increase of heat when the conductor is greatly reduced in diameter. This can be seen by examining the electric light in our houses. The main conductors are of large size and of pure copper, while the filament of the lamps is fine, and is of carbon. The narrowing of the electric conductor, therefore, leads to a great development of heat and light; and here we must bear in mind that the only difference between heat and light consists in wave length; the heat waves are much longer than the light waves. As we continue our study we shall find also that the only difference between light, heat, and electricity is in the length of waves.

Our analogies between pipes conveying horse power by means of steam or compressed air and wires carrying electricity, we have seen, do not lead us far. There is an evidence of pressure in such pipes, and also on conductors carrying electric currents. There is a development of heat on both pipes and on electric conductors, but this development is much greater in the case of electricity. Here our analogy stops short. The steam pipe exerts no influence outside itself to attract or repel other pipes. Its effect on a magnet is the same whether the steam flows through it or not; it acts in both cases like a piece of iron; it is not aroused to the exertion of a tremendous power on a neighbouring magnet or on a neighbouring conductor carrying a current when steam rushes through it; it is dead to changing magnetic influences. But the electrical current in a conductor seems to exert a mysterious influence on all neighbouring objects, even on the surrounding air. This influence is especially marked on magnets and on neighbouring electric currents. The western end of the Jefferson Physical Laboratory of Harvard University was constructed with especial reference to freedom from magnetic disturbances, in order that delicate experiments in electricity and magnetism might be conducted there. All the steam pipes and gas pipes were constructed of brass, and no iron nails were used in the flooring. This construction was costly; but it has been made useless by the presence of the overhead wire of an electric road. Every time an electric car passes a fluctuation of the electric current is caused on the overhead wire; all the delicate magnetic instruments in the laboratory throb in unison with the electric disturbance, and this, too, at a distance of at least 400 feet. In scientifically enlightened Germany the electric cars are not permitted to run near the great physical laboratory at Charlottenburg, near Berlin.

We see, therefore, that the mysterious phenomenon we call an electric current is extremely versatile. It more nearly resembles the nervous action of man than any other influence with which we have to deal. We can not, however, speak of a nerve current, and we shall see later that strictly considered we ought not to speak of an electric current, for there is but little evidence that there is a flow of electricity in a wire which we ordinarily say conveys a current. I shall continue to use the term electric current only in a popular sense.

Before we leave our consideration of what may be called the palpable phenomena of the electric current, and before we enter upon a study of the source and generation of this current in a conductor, let us dwell a little longer on the development of heat produced by electricity on conductors. In our cities it often happens that the overhead wire of the electric railroads falls to the ground. When it touches the rails the wire is raised to a white heat, and contorts and twists like a fiery serpent. With a stick or a thickly folded newspaper it is perfectly safe to lift the wire from the rails and thus stop the development of heat. If it were not for the mere heat, one could take hold of the wire with a handkerchief and run no danger of an electric shock; for we have said that a piece of paper is sufficient to stop the flow of an electric current which represents many thousand horse power. A general knowledge of the impunity with which we can handle wires carrying strong currents, if we seize them properly—that is, with a folded handkerchief or with a folded newspaper—would often prevent needless terror and save life. From our consideration, therefore, of the development of electric light when a conductor is reduced to the size of a thread, and from the study of the phenomenon of the great heating of the overhead wire of an electric road when it falls to the ground and touches the rails, thus gaining a quick return to the power house where the source of electricity is situated, we are led to two conclusions: First, that a narrowing or constricting of the conductor, or, in other words, an increase of resistance, leads to a great development of heat and light; second, that a shortening of the path or a lessening of resistance also leads to a great development of heat. These are apparently contradictory conclusions. The contradiction, however, is explained when we discover by suitable measuring instruments that, whenever we diminish the diameter of an electric conductor in order to produce light or heat, we reduce the total flow of electricity through the entire electric circuit; less goes in general through the large wires leading to our electric lamps when they are lighted than when they are not lighted. The heating of an electric conductor is proportional to the square of the strength of the current which is flowing; that is, if the strength of the current is doubled, the amount of heat developed on the conductor is quadrupled. We can see, therefore, why the large overhead wire of an electric railroad is intensely heated when it falls from its guard wires and touches the rails of the road. The current does not encounter the resistance of the overhead wire beyond the point where it touches the rails, as it usually does when it is in its normal position. This portion is cut out by the fall of the wire, and the entire current rushes through the wire between the feeding wires and the spot where the broken wire touches the ground and forms a powerful electric arc at the latter place. If one should notice the character of the overhead wire now used on electric railroads, it will be found that it is of copper of nearly the diameter of a small lead pencil. Such a wire opposes very little resistance per foot to the flow of the electric current which is used to propel the cars. Its resistance, however, becomes very appreciable over long distances, and this resistance is one of the reasons why electrical power can not be transmitted economically farther than five or six miles by a steady current. In order to increase this distance we must increase the size of our copper conductor to diminish the resistance, and the cost of copper speedily sets a limit to our endeavours. We shall see, however, as we continue our study, that with a to-and-fro or alternating current of electricity—that is, an unsteady current—we can use a smaller copper wire, and thus transmit electrical power over hundreds of miles.

The resistance of copper and silver to the steady electric current is less than that of any other metals. The resistance of iron is at least six times that of copper. A very slight impurity in copper wire exerts a marked increase of resistance. The use of a 98-per-cent copper wire in the place of a 97-per-cent copper wire on a long-distance telephone circuit between Boston and Chicago would mean a great saving of copper, and therefore of money. Twenty-five years ago any specimen of copper wire would vary in resistance in different portions of its length by at least 1 per cent, and sometimes more. To-day it is difficult to detect any variation in resistance in hundreds of feet of copper wire which is furnished by manufacturers. This perfection of material is due to the demand for good conductors by electricians. The production of copper has now become of more importance to the world than the production of silver; this is largely due to the remarkable practical developments of electricity.

An interesting application of the phenomenon of heating due to increased electrical resistance is made in surgery. An electrical current is conducted through a fine platinum wire which is raised to a white heat by the current. This localized heat can be used to cauterize portions of the body to which heat could not hitherto be applied; for instance, in the throat and in the nasal passages. On certain electric roads, also, the cars are heated by coils of iron wire through which an electric current passes. The heat is due to the resistance of the iron wire. It is thought, indeed, by some that all the culinary operations in our houses in time will be performed by electricity. On a gridiron of wire rendered red-hot by an electric current a beefsteak can be readily cooked. Water can be heated in receptacles in which are placed coils of iron wire; and the great advantage of such applications of electricity to the operations of domestic cookery are similar to those in its applications to surgery: the heat can be applied exactly at the spot where it is most needed.

When the dynamo machine became an efficient source of electrical currents men's minds were immediately turned to the problem of what was called the subdivision of the electric light, or, in more appropriate words, the problem of obtaining a number of small incandescent lights from a given constant electro-motive force and a given current. It was an easy matter to obtain an arc light between carbon points, but this light was suitable only for street lighting, and could not be made to contribute to the comfort of humanity in each home. It was early recognised that whoever could aid in solving this great problem would render mankind a service. As we have said, many minds were at work upon this problem and partial solutions were obtained. It was soon perceived that the small electric light, if it were invented, must be obtained by raising a fine metallic wire or a carbon filament to a high state of incandescence. At first, fine platinum wire was selected for this purpose, since its high degree of infusibility seemed to render it suitable; it would not burn out, so to speak, under the electric heating. It was found, however, that its state of incandescence was variable, and that even when the platinum loop was placed in a little globe from which the air was exhausted this inconstancy of action manifested itself, and no reliance could be placed upon the platinum-incandescent electric lamp.

When, however, a suitable filament of carbon was substituted for the platinum wire in a globe from which the air had been exhausted, a lamp was obtained which would burn from three hundred to six hundred hours. To maintain a number of such lamps on an electric circuit, it was speedily discovered that they must all receive the same electric pressure, so to speak—that is, the current must be urged through them by the same electro-motive force. Moreover, each lamp required the same amount of current to raise it under the given electric pressure to the same candle power. This could not be accomplished by placing the lamps one after another on the same wire, for the resistance offered by each lamp diminished the current. Two or three lamps placed in this manner were sufficient to prevent a great dynamo, capable of manifesting an energy equivalent to many hundred horse power, from exhibiting this energy along the electric circuit. A simple application of Kirchhoff's laws soon solved this difficulty. If many paths are offered to the electric current, it divides itself in a proportional manner among these paths; for instance, if we should consider the meridian lines of the earth as electric wires meeting at the poles, and we should lead an electric current into these wires at the south pole of the earth and out at the north pole, the current would divide itself proportionately on the meridian wires. Each meridian wire would receive the same amount of current, and each subdivision of current would be under the same difference of electric pressure —namely, the difference of pressure at the two poles. A little lamp, therefore, placed on any wire represented by a meridian would glow with the same brilliancy as that of its neighbour on another meridian; and, furthermore, according to Kirchhoff's laws, the resistance opposed to the electric flow by the lamps would be enormously diminished by this arrangement, which is termed the multiple circuit. We can obtain a conception of a multiple circuit and its effect in diminishing resistance by considering the flow of water through pipes. If we narrow the diameter of a pipe we increase the resistance to the flow of water; if, however, we should undertake to carry water from one pole of the earth to the other pole by means of pipes, we could accomplish this with the least resistance by employing a number of small pipes of small diameter, situated like meridians of the earth, all connecting with the principal mains at the north and south poles. The greater the number of small meridian pipes the less the resistance to the flow through the large mains.

By means of the carbon filament in an exhausted globe and by the arrangement of the multiple circuit the incandescent system of electric lighting became a commercial success.

CHAPTER V.

FLOW OF ELECTRICITY IN THE EARTH.

In the following chapter let us examine the passage of electricity through the earth; for it is well known that it was discovered in the early days of telegraphy that a return wire between Boston and New York, for instance, could be dispensed with, and that the earth could be used instead of the return wire, thus halving the injurious resistance of the circuit; for it was found that the earth did not oppose any appreciable resistance compared with the total length of the telegraphic circuit.

We can find no analogy between the flow of steam, gas, or water and the case of the return circuit through the earth. In the case of steam, gas, and water, and of all fluids forced through pipes from a power-house, nothing returns to the power house if we should connect the pipes to the ground; for the steam would be condensed, the air pressure lost, and the water would soak into the ground. In the case of electricity, however, nothing is lost by connecting the wires leading from the power house or battery to the ground. Indeed, in certain cases a great deal is saved, for the energy of the current is not dissipated into heat along a return wire. We have said that a magnet or compass needle instantly points to a wire through which an electrical current is passing. It is like the finger of a mute person pointing out a secret. It points to the wire if it is moved along the wire from one earth plate to which the wire may be attached to the power house or battery, and from the power house or battery to the earth plate at the other end of the wire circuit. If placed on the earth between these earth plates and sufficiently far from the overhead wire—on the ground, for instance, beneath an ordinary telegraph wire strung on poles—the compass or magnet is quiescent and performs its normal task of pointing to the poles of the earth. It gives no evidence of an electric current in the ground; the electricity, so to speak, seems to have leaked away like water. Yet instruments show that the current apparently flows out from the power house in one direction to the ground and returns from the ground to the power house.

If we should take a minature earth—a globe of metal, for instance—several feet in diameter, and run an electric current to what may be called the north pole of such a globe and lead it away from the south pole, we shall find that the current apparently spreads out from the north pole and converges, so to speak, to the south pole. If the globe were 20 feet in diameter very little indication of a current would be obtained around the equator of such a globe. Let us now build up a globe made of steam or water pipes all connected to one main pipe at the north pole, and again at the south pole. We can suppose the pipes to represent the divisions of an orange. When steam or water leaves the main pipe and is divided in its flow equally among the pipes, placed similarly to the divisions of the orange; the amount of flow through any one pipe can be made very small, although the flow through the main pipe leading to the globe is very large. If we should connect any two neighbouring pipes along the equator so that water or steam could flow from one to the other, we should find that there would be no flow, for the pressure at the two ends of such a connecting pipe is the same; there is no difference of pressure to force the steam or water from one pipe to the other. If, however, we should connect one pipe at a point on the equator—in Africa, for instance—with another pipe at a point corresponding on the globe to New York, there would be a flow in the connecting pipe, for there would be a difference of pressure. In the case of electricity, a telephone will determine whether there is a flow from one portion of the earth's surface to another when we lead an electric current into the earth and out of it. Let us use the telephone at first merely as a detector of an electrical flow, just as we used in the above illustration a pipe connecting two pipes in order to determine whether there is any possibility of a flow of water between them. That it can be so used we can easily ascertain, for we hear a click in the telephone whenever we touch the two wires leading to it to the two poles of an ordinary battery such as is used on bell wires or for medical purposes. If we should hold a telephone to the ear and connect one of its leading wires to the rail of an ordinary electric road and the other to the iron posts which run beside the track, we should hear a click at the moment of making the contact with the iron pole if there is a leakage of electricity from the overhead wire, which is supported by the iron pole and its connections, into the ground. In other words, a difference of electrical level would be shown between the iron post where it enters the ground and the rail. If now we should make a globe of a number of copper wires, insulated from each other and forming the meridians of such a globe, and connect all these great circles of copper wire together to one wire at the north pole and to another wire at the south pole, and lead a current of electricity into the collection of meridian wires at the north pole and out of the collection at the south pole, we should find that very little current would go through any one wire; and if we should connect our telephone wires with two neighbouring wires anywhere along the equator we should hear no click; there is no flow of electricity between points of the same pressure. If, however, we connect one wire of the telephone at one point on the equator and the other wire at a point on the wire globe corresponding to New York, we should hear a click, for there would be a flow between these points.

From such experiments we see that what we call the electric current flows out in all directions from the point where it enters the earth, and appears to converge again to the point where it leaves the ground to enter the wire and to return to the power house or battery. Perhaps the best illustration of the manner in which the electric current spreads out in the earth is afforded by a method of telegraphing without wires, which I described in the Proceedings of the American Academy of Arts and Sciences, and which has lately been repeated by Prof. Rubens in Berlin, and by Mr. Preece of the London telegraphic system.* In my paper I remarked: "The theoretical possibility of telegraphing across large bodies

* My original researches were made between the observatory at Cambridge and the city of Boston, which were connected by a time-signal wire. The current upon this wire was broken by a clock at regular intervals. I found that I could hear the clock-beats a mile away from the wire by connecting a telephone to a wire and by grounding the ends of the wire 500 or 600 feet apart and parallel with the time circuit.

of water is evident from this survey which I have undertaken. It is possible to telegraph across the Atlantic Ocean without a cable. Powerful dynamo-electric machines could be placed at some point in Nova Scotia, having one end of their circuit grounded in Florida, with an overhead wire between these points of great conductivity and carefully insulated from the earth except at the two grounds. By exploring the coast of France two points not at the same potential could be found, and by means of a telephone of low resistance the Morse signals sent from Nova Scotia to Florida could be heard in France."

What we have said in regard to the spreading out of the electric effect or current in the earth is entirely applicable to the case of the human body. If one pole of a battery or other source of electricity is applied to the middle of the back and the other pole to the middle of the breast, the electric current which is thus led into the body spreads out like a stream of water through an infinite number of fine holes in a rose jet; it permeates every muscle in a greater or less degree between the back and the breast. Its flow can not be detected in the body by a telephone, but a delicate galvanometer, which is the electrician's microscope, will show its spreading. Not only can the spreading of the current be detected by galvanometers, but this action can also be studied by chemical analysis, as we shall see when we study the passage of electricity through fluids. Before going further in the subject of the earth circuit we can already perceive that the use of the earth for a return is not always desirable, for in the neighbourhood of a great city the earth becomes filled, so to speak, with the electrical flow from the common use of the earth by the telegraph companies. At one time, as I have shown in the article already referred to, it was possible to adjust one's watch by connecting a telephone to the water pipes and gas pipes in almost any part of Boston and Cambridge, for one could hear the clicks of the observatory clock from which time signals were sent. The telephone companies no longer, however, use the earth for a return circuit on their long-distance lines, and employ an entire metallic circuit of copper wire. This circuit obviates the earth disturbances due to the spreading out of electric circuits; it has also other advantages, which we will study later. It is interesting to observe here that what was once considered a notable practical advantage in telegraphy is fast losing its importance as we refine upon our methods of transmitting intelligence by electricity. We shall also see that the earth is no longer used by the electric light and power companies. We shall see further that what we call the steady current is being replaced by the unsteady current or the to-and-fro current for the electrical transmission of power over great distances; and, still stranger, we shall perceive that there are reasons for believing that there is no electric current or flow of electric energy on the wires which are conveying telegraphic messages or propelling electric cars; and that for very rapid alternating currents copper is really a poor conductor, and glass an excellent one.

There are several terms now in common use in electrical science which serve as measures of value, and I shall endeavour to give a popular explanation of them. The term ampère is used to denote the strength of the electric current; the word volt, to denote the unit of electro-motive force or electrical pressure on the circuit. The current may be said to flow under a head, which is termed the voltage. This head is analogous to a head of water which forces a current of water through a pipe. The quantity of water which flows through the pipe in a unit of time—say a second—is a measure of the flow of water; the quantity of electricity which flows in a second of time is a measure of the electrical flow, and is called an ampère. This flow meets with a certain electrical resistance, which is termed an ohm. The flow of water, also, through a pipe meets with a resistance in the friction with the pipe. These terms—ampère, volt, and ohm—have passed into daily use. They perpetuate the names of a great Frenchman, a great Italian, and a great German. There are two other terms, not so readily comprehended by fluid analogies: The farad, named for Faraday, the unit of electrical capacity, the unit of electrical quantity we can store up; and the henry, named for the great American Joseph Henry, the unit of inductance or electrical inertia—an inertia which manifests itself when a current suddenly rises or falls. These two terms—farad and henry—have immense importance in the subject of alternating currents of electricity.

CHAPTER VI.

THE VOLTAIC CELL.

It is interesting to reflect that the study of electricity received its greatest primal impulses from two nations so unlike in mental characteristics—the Anglo-Saxon and the Italian. To Benjamin Franklin we owe a clearer conception of the phenomenon of lightning, and to Galvani and Volta is due the discovery of the electrical battery. In later times we are indebted to the Anglo-Saxon for the discovery of the principle underlying the action of the dynamo machine and the telephone. The story of the discovery of the electric battery is well known, but we will repeat it for the sake of pointing out modern interpretations of the mysterious action which puzzled Galvani and Volta. In a small cabinet of the Jefferson Physical Laboratory of Harvard University there are three instruments which represent all that was known about electricity in 1830. There is a Franklin electrical machine of great size, with its glass globe and its rubbers, and its ponderous wheel for turning the globe against the rubbers, ordered by Benjamin Franklin for the College at Cambridge; there is a voltaic battery consisting of a great many zinc plates and copper plates which can be immersed in a suitable acid; finally, there is a large electro-magnet, simply a horse-shoe-shaped piece of iron wound with coarse wire. The electrical machine and the battery have become antiquated; there is nothing in that case of practical use save the electro-magnet. Moreover, the distinction between the electrical machine and the battery has disappeared. We shall see that the electrical machine is simply a battery, and that when Benjamin Franklin rubbed a glass rod with a piece of catskin he was dealing with the phenomenon which takes place in any voltaic cell.

In 1791, just as Benjamin Franklin was passing off the stage, Galvani made the observations which led to the voltaic cell, and directed philosophers' contemplation from the clouds to the phenomena on the earth and in its substances, so to speak. To Galvani we owe a great debt, greater far, I believe, than we owe to Franklin, for Galvani opened a vast field of inquiry which led to the discovery of electro-magnetism, and gave Faraday the means of discovering induction. Before Galvani's time men were lost in philosophical speculations in regard to subtle fluids. After his experiments their thoughts were directed to the conditions of matter immediately about them. Benjamin Franklin brought electricity down to earth from the clouds, while Galvani's experiments brought men's minds down from the heights where they were lost, having no tangible transformations to study. His own account of the beginning of his experiments is extremely interesting, for it shows that he was not anxious to make it appear that he was the first to notice the strange phenomena which proved to have such far-reaching results. He says:

"This is the way the thing happened. I dissected a frog and prepared it as is shown in Fig. 6, and laid it on a table upon which stood an electrical machine (Fig. 6), far from the prime conductor and not in a straight line with it. When one of the servants, who was at hand, touched with the point of the dissecting knife the inner lumbar nerve (D D) of the frog, all the muscles of the thighs appeared to contract as if under the influence of powerful cramps. The assistant thought that the phenomenon occurred when a spark passed between the conductors of the electrical machine. Astonished by this new phenomenon, he turned to me, I being occupied in other matters and absorbed in thought. Thereupon I was inflamed by an incredible haste and desire to prove the same, and bring the hidden mystery to light." *

Here, with extreme candidness and generosity, he gives full credit to his assistants for the accidental discovery. He traced the cause of the strange convulsions to the working of the electrical machine, and found that discharges of lightning could also produce the movements of the muscles of the frog's legs. He continues thus:

"After I had investigated the effects of atmospheric electricity my heart burned with desire to test the power of the daily quiet charge of electricity in the atmosphere." He had noticed that the prepared frog's legs, bound up with brass hooks on an iron railing, showed the same contractions not only in the case of thunderstorms, but also under a clear sky. He was thereupon led to study the effect of touching the nerves with different metals and with non-conductors, such as glass and wax. He found that when the circuit between the nerves was made by two different metals powerful contractions ensued.

---

* De Bononiensi Scientiarum et Artium Instituto atque Academia Commentarii, tomus vii.

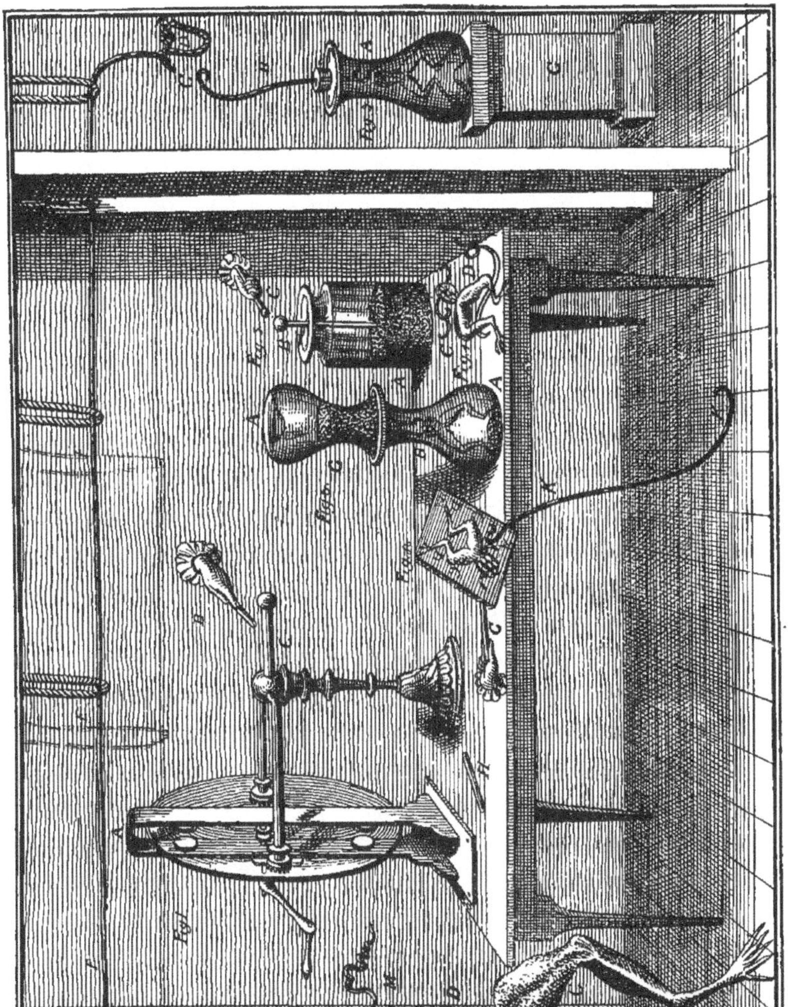

In Fig. 6 we see a representation of the various experiments which he tried. The twitching of the frog's legs served him for a galvanometer—a sensitive indicator of the electrical current which was excited in the circuit of the metals and the muscles and nerves of the frog. A new instrument in physical science often opens a great field of discovery. The frog's legs in the hands of Galvani and his co-workers proved to be such an instrument.

It was not a difficult step to take from the standpoint of Galvani to that of Volta. The instrument was at hand and the phenomenon had been observed. Galvani attributed the action to the vital electricity of the nerves and the muscles of the frog, while Volta attributed the action entirely to the contact action at the junction between the two metals. The controversy as to the cause and seat of the electro-motive force between two metals in an ordinary battery is still a matter of dispute, and we are little wiser than philosophers were in the days of Galvani and Volta. The labours of these two men, however, opened, as we have said, a great field in electricity. Men began to study the manifold phenomena of the electric current produced by batteries, and the next great step was made by Oersted, who discovered the principle of a new instrument, the galvanometer, the indications of which have led the way to the great practical employment of electricity.

Galvani, we have said, attributed the source of the electrical current which convulsed the frog's leg to the animal electricity of the frog. Volta, however, soon showed that an electrical effect could be obtained without the use of the frog's leg by connecting a piece of zinc to a piece of copper, and he was led to the invention of the voltaic pile, which in its first form consisted merely of alternate discs of copper and zinc separated by pieces of blotting paper moistened with salt and water. Volta attributed the effect observed by Galvani to the contact between the two metals which were employed to touch the frog's legs. In an ordinary battery the plates forming the battery are not in contact, but are separated by a layer of liquid, which acts chemically upon one of the plates. The question of the seat of the electro-motive force in a voltaic cell is as much a mystery to-day as it was in the times of Galvani and Volta, but we are beginning to see that our best hope of solving the mystery consists in studying the transformations of energy in the battery, and in measuring the heat developed under different conditions. The chemical action is an evidence of an increased molecular activity, so to speak, and in general terms we can conclude that whenever in a circuit we have a difference of molecular activity, at two points in the circuit an electrical circuit results. The phenomena which takes place in a battery consisting of two metals with a liquid between the two metals appear to be far more complicated than those which are manifested along the wire connecting the metals or along the outer circuit. The liquid is broken up. If it is acidulated water, oxygen is given off at the positive pole and hydrogen at the negative.

It can be said, in general, that an electrical current is generated whenever two dissimilar metals connected by a wire are immersed in a liquid which is capable of conducting electricity. For instance, if the handles of a silver spoon and an iron spoon are connected by a copper wire, and the bowls of the spoon are immersed in a tumbler of salt and water, an electric current passes from the silver to the iron along the copper wire, and in the water from the iron spoon to the silver spoon.

Such a battery would be sufficient to send a signal under the Atlantic from America to England, but it would not be powerful enough for commercial use. Before the iron spoon is connected with the silver spoon by a wire, very delicate instruments will show a negative charge of electricity upon the iron spoon, and a positive charge upon the silver spoon. By the use of a great number of zinc and copper plates separated by paper moistened with salt and water, Volta was able to show that a body charged with positive electricity was repelled by the terminal connected to the copper plates, and attracted by that connected to the zinc plates. Indeed, by greatly increasing the number of tumblers containing plates of copper and zinc immersed in salt and water, and connecting each copper plate to each zinc plate by a wire, one can obtain electrical sparks when the final copper plate is brought near to the first zinc plate.

Since any two metals immersed in a suitable conducting liquid constitutes a voltaic cell or battery, it will be readily seen that the number of forms of such batteries is very great. The Leclanché cell, which is so commonly employed at present in houses to supply the electric current for bells, consists mainly of a rod of zinc and a rod of carbon, both immersed in a solution of sal ammoniac. When Prof. Tyndall came to America to deliver lectures on physical science, he brought with him a hundred Grove cells, which consist of platinum plates immersed in strong nitric acid which is contained in a porous cup; this cup is placed in another receptacle filled with sulphuric acid and water, in which there is a zinc plate. The electrical current flows from the platinum to the zinc outside the cell, and from the zinc to the platinum inside the cell. The cells are joined, as we have said, in series, zincs to platinums, and finally the last zinc is joined to the first platinum. By means of fifty such cells Prof. Tyndall showed the electric light, and illustrated the subject of optics by its aid. The labour of preparing such a battery for each lecture was very great, the nitrous acid fumes were dangerous, and the battery would not furnish a current sufficient for the production of light equal to an ordinary street electric lamp for more than an hour. The lecturer on physics to-day finds a source of electricity on hand similar in abundance and ease of application to the supply of gas or water. This supply of electricity, however, is obtained not from batteries but from dynamo machines. We shall describe the dynamo in a subsequent chapter, and point out what a revolution it has accomplished in the practical uses of electricity. The voltaic cell, however, can never lose its scientific importance, although it has ceased to be looked upon as a commercial source of electricity for lighting purposes or for the transmission of power. Why two different metals immersed in a conducting liquid produce a current of electricity is not well understood. This question was the cause of a memorable discussion between Volta and Galvani, as we have seen, and within a few years the British Association appointed a committee to collect and classify the investigations on the seat of the electro-motive force in a voltaic cell.

I have said that the battery is still of great scientific importance. We are brought face to face in its action with the great problem of the connection of electrical action and molecular movement; indeed, with the fundamental principles of chemical action. Modern chemistry is rapidly becoming a study of motion, and this can also be said of electricity. We have, then, a common ground of attack in endeavouring to penetrate into the mysteries of chemical action and electrical action. When we touch the lower surface of our tongue with the wire from one plate or pole of a battery, and the upper surface with the wire leading to the other pole of the battery, we can taste, so to speak, the electric current. We are conscious of a peculiar tingling sensation in the tongue. If we lead the wire into acidulated water, we find that the water is broken up into its constituents—bubbles of hydrogen gas are given off at one end of the wire immersed in the water, and bubbles of oxygen at the end of the other wire. This action is called electrolysis, and we taste it, so to speak, when we touch our tongue, with its tissues rich in fluids, to the wire terminals. Instead of the sensation of taste, it is rather a slight electric spark which we feel, which arises from the electro-motive force which produces electrolysis. The action of electrolysis is immediately connected with molecular motion. The electrical action has rent asunder the bonds that connected the hydrogen molecule to the oxygen molecule—the bond that made up the particles of water—and these hydrogen and oxygen molecules are now free to vibrate independently of each other, and to enter into new combinations. We shall find that their mere adherence to the ends of the wires dipping into the acidulated water is sufficient to produce an electric current; for if we remove the battery which has broken up the water into its constituents, and connect together the ends of the wire which were connected to the poles of the battery, we shall obtain again an electrical current. Furthermore, we find that by means of electrical action we can transport a substance through the tissues of the human body—for instance, from the back to the breast of a man.

The electrical current, therefore, flowing through a conducting fluid, exerts a marked effect upon the molecules of the fluid. In this respect its behaviour in passing through a fluid differs greatly from that which it manifests in a wire or metal conductor. No molecular effect has yet been observed in wires through which currents of electricity have passed, save the effect due to the heating of the wires. Since the earth is composed of both solid and liquid matter, we should expect to find some evidence of chemical action set up by the spreading of the electric current. This evidence is very strong in the case of electric railroads. The current used by such roads returns to the power house by the rails of the road, by conductors connected with these rails, and by the earth. The current spreads out in the earth, and seeks the shortest passage back to the power house. It has been found that the gas and water pipes of a city in which an electric road is situated exhibit a chemical action due to the electrical current of the railroad. This action is similar to that we observe on the ends of wires which lead an electrical current to and away from a conducting liquid. For instance, bubbles of oxygen adhere to the end of the wire by which the current enters a tumbler of salt and water, and bubbles of hydrogen to the end by which the current leaves the tumbler. This action is called electrolytic polarization. It was found in Boston lately that the lead pipes in certain localities were filled with minute pinholes where this electrolytic action had eaten away the lead. This action has been lessened by connecting the positive pole of the source of electricity—the dynamo—with the rails, and the underground pipes and the negative pole with the overhead wire, so that the current of electricity should flow from the underground pipes, the earth, and the rails through the wheels of each car to the trolley, and back to the power station along the overhead wire.

In this way the electrolytic action is exerted less strongly on the underground pipes.

The most striking example of this action of the current in passing through a liquid is in the case of what is called the storage battery. If, instead of leading a current of electricity into a tumbler of water and out of it by means of copper wires, we should connect the leading-in wire to a strip of lead, and the outgoing wire also to a strip of lead—the lead strips being immersed in the water in the tumbler—we should have an elementary modern lead storage cell. Bubbles of hydrogen adhere to one lead strip, and bubbles of oxygen to the other, when a sufficiently strong electric current is passed through the water in the tumbler. These gases arise from the decomposition of the water by the electric current, and this decomposition, we have said, is called electrolysis. The amount of oxygen and hydrogen adhering to the lead strips can be enormously increased by covering the lead strips with oxide of lead. The oxygen given off on one lead strip is held there in great quantity by the conversion of the oxide of lead into a higher oxide, that is, an oxide with a larger amount of oxygen—the peroxide of lead—while the hydrogen gas lowers the oxide of lead on the other lead strip to a finely divided porous layer of hydrogenated lead. When such lead strips, after having been charged by an electric current, are connected by an external wire, a strong current rushes from the oxygenated lead strip to the hydrogenated strip. By increasing the surface of such strips we can charge such a storage battery to an enormous amount. If Prof. Tyndall could have had a storage battery of fifty cells instead of fifty pint Grove cells, the charging of such a battery would have sufficed for an entire course of lectures, and the labour of preparation would have been greatly diminished. It will be observed that the term storage battery is a misleading one, if we understand by the term storage a storage of electricity. What is really stored is chemical action due to the electrical current.

One of the chief practical objections to the employment of the lead storage battery is its weight. A lead battery sufficiently powerful to propel a bicycle or tricycle, developing one quarter of a horse power, would weigh at least 300 pounds, and it would not furnish power for more than four hours before it would need to be recharged. Prof. Langley informs me that he has constructed a little steam engine which weighs less than two pounds and which develops one quarter of a horse power. Some of the recent forms of petroleum motors for horseless carriages develop at least two horse power and weigh only fifteen pounds. It will be seen, therefore, that the storage battery in its present best form, which is the lead form, is not of much practical importance in the subject of the transmission of power; it is, however, of the greatest theoretical interest.

With five thousand little lead cells charged by a dynamo machine sparks can be obtained, and a severe electric shock is felt when the positive and negative terminals of the battery are seized by the hands. The battery imitates in all respects the action of Franklin's electrical machine; and it is interesting to observe that at first the celebrated experiment of Galvani seemed to lead us away from the study of frictional electricity, which had exclusively filled philosophers' minds before the date of Galvani's experiments. Our increase of knowledge, however, of galvanism has conducted us back to the field in which Benjamin Franklin worked, giving us a little more light upon the subject of what is called frictional electricity. At first sight nothing seems more remote from the phenomena produced by rubbing a cat's fur, or the phenomenon of lightning, than the action of the battery which rings bells, decomposes liquids, and produces the mysterious magnetic effect in the neighbourhood of wires which we have noted. We have now good reasons for believing that when we stroke the fur of a cat the mechanical action breaks up the arrangement of the molecules of the hair and of the surface of the hand, just as the electro-motive force of a battery breaks up the arrangement of molecules in a conducting fluid and produces an electric charge on the two metals of the battery. When a lady produces a spark by walking across a carpeted floor, the molecules of her silk skirts and dry garments are rudely disarranged and an electric charge results. This charge does not result from any peculiar electricity of the body. It is not what is termed animal electricity. It results merely from the rubbing of the clothes or of the dry slippers upon the carpet, and it can be produced by men as well as women, being merely a question of the proper clothing necessary for its production.

In our furnace-heated houses in winter the phenomena of electrical charges produced by friction is of common occurrence. A pair of silk undergarments suddenly withdrawn from a pair of trousers diverge under their electrical charge. A sheet of paper briskly rubbed adheres to objects presented to it. The sheet of paper, together with the object which is brought near it, really constitute a battery—the sheet being one pole and the neighbouring object being the other pole, while the air between takes the place of the fluid in the ordinary battery. The Franklin electrical machine can be considered a battery, and can be made to produce all the effects produced by an ordinary voltaic cell. We shall see, when we consider the dynamo, that the latter also can produce all the effects due to batteries, and it can be made to give sparks five feet long which are identical with discharges of lightning.

In general, a change of molecular aggregation produces a manifestation of electricity, and when we reflect upon this fact we see the immense importance to the chemist of the study of electricity. If we should suddenly bend a ring of metal wire we can produce a current of electricity in the wire. Moreover, if we hold it over a lamp and heat it at one point we can also produce a current in it. In these cases also the current is produced by the change in molecular arrangements in the wire ring. We are apt to hastily conclude, from observations upon the electricity produced by differences in molecular activity, that electricity is a motion or vibration of molecules. As we continue, however, our study of what electricity is, we perceive that there are other and more significant ways of producing electricity than by chemical action, or by any operation which breaks up and changes the arrangement of molecules.

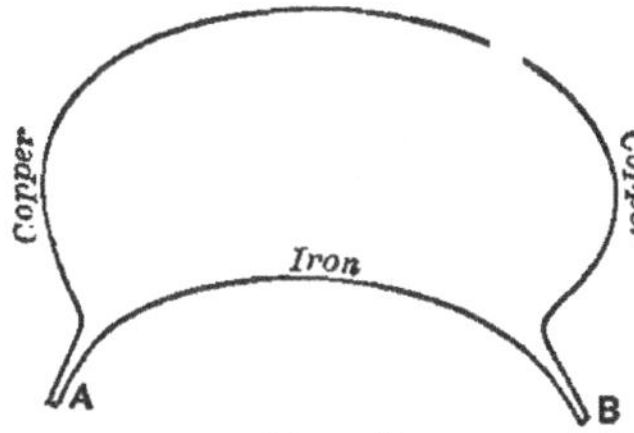

Fig. 7.

A voltaic cell is not our only means of obtaining a current of electricity without the use of a dynamo. One of the most valuable means for an experimenter consists in the employment of heat. If we join any two metals—for instance, iron and copper—and heat one junction—for instance, A, Fig. 7—and cool the other, B, an electric current results which flows from the hot junction through the copper wire to the cold junction. The new alloy termed constantan, with copper, gives a very sensitive combination. These thermo-electric junctions can be used as very delicate thermometers. One can easily determine the one hundredth of a degree Centigrade by means of them. To detect the current, it is merely necessary to have a delicate galvanometer. Prof. Tyndall, in his remarkable treatise on Heat as a Mode of Motion, devotes much space to the description of his thermopile and the galvanometer he employed, for this apparatus served to illustrate throughout his treatise the various manifestations of heat due to motion. He used it instead of a thermometer. Since the date of the publication of Prof. Tyndall's treatise, which gave to the world in a popular form the great generalization of the conservation of energy, this form of electrical thermometer has been much simplified and made more sensitive. Instead of the needle galvanometer employed by Prof. Tyndall, we now have the mirror galvanometer, and we are able to detect electrical currents which his instrument would not respond to.

Thermal currents always arise when there is a difference of temperature between any two points in an electric circuit. It is not necessary to employ the junction of the different metals to produce these currents. If a piece of copper, for instance, is heated at one place and cooled at another, a thermo-electric current results. If a knot is tied in it and it is heated on one side of the knot, a current results. In other words, any change in the molecular aggregation of the metal at the points will produce a current of electricity. An attempt has been made to apply this principle to practical use; for instance, a furnace has been constructed with thermal junctions set in its walls, with the other set of junctions outside. It is possible, by having a large number of junctions in the pot of a furnace, to produce an electric light. The heat of an ordinary house furnace is sufficient to afford a supply of electricity for the house bells and for running a sewing-machine, and, indeed, for a certain amount of electric lighting; but no practical way has been devised of preserving the junctions from injury due to expansions and contractions. The energy in the coal we use in ordinary furnaces is amply sufficient both to heat and light our houses if it could be economically transformed. It has been proposed by various investigators to employ the electro-motive force developed between iron and carbon which are immersed in a hot alkali. The carbon wastes away or is consumed under the action of a current of air which is forced through the melted alkali. If we could break up the carbon into its constituents, or ions, as we can water, we could produce electricity directly from coal. The use of the thermal junction seems at present the only way worthy of consideration by means of which we can produce electricity direct from coal without the use of a steam engine. The electro-motive force produced by heating the junction of two metals, however, is very small even with the best combination—bismuth and antimony—far less than with the employment of copper and zinc in the ordinary voltaic cell.

Both with the employment of two different metals and in the case of an electrolyte, outside the evidence of the increased molecular activity at the surface of the metals and an increase of heat throughout the liquid of the cell, there is no evidence of any flow in one direction or the other, or of any commotion in the liquid. At a sufficient distance from the plate of the battery the liquid is in repose. A beam of light sent through it is not absorbed by the liquid more or less when the current flows. A photograph taken by means of light which has passed through a layer of liquid traversed by an electric current differs in no respect from one taken through an equal layer of the liquid not traversed by a current. No effect of strain can be observed in the liquid. Faraday examined this question carefully by means of the polarization of light. There are certain substances which confine the vibrations of light in one plane. We can think of such substances as similar in internal arrangement to a Venetian blind with its horizontal slats. If we should attempt to thrust a hand moving to and fro through the blind, it would be possible to do so only when the hand is moving to and fro horizontally. Now the vibrations of a ray of light are in all directions, so that a section of the ray would be a star-shaped figure. Only those vibrations which are in the horizontal plane would therefore pass through the blind. When this light emerges from such an arrangement of matter, therefore, it is said to be polarized, and it can be tested by another substance exactly similar to that which polarizes the ray; for if the analyzer is turned so that its molecular arrangement, in regard, for instance, to a horizontal plane, is the same as the polarizer—that is, if two Venetian blinds are hung vertically one behind the other—the hand moving horizontally can be made to pass through both. If one blind, however, is not vertical, and the slats are at right angles with the other, the movements of the hand will be intercepted. Two Nicol prisms form such an arrangement. Although they are perfectly limpid, like two pieces of glass, yet when we look through both and turn one we shall find that all the light is cut off in a certain position of the prism next the eye, which is called the analyzer. When a piece of glass submitted to a strain is placed between the two Nicols, we find that we are obliged to turn the analyzer through an angle to obtain extinction of the light, which is different from the angle of extinction when the glass is not under strain. Polarized light, therefore, enables us to detect a strain. Faraday could not, by an arrangement similar to that we have described, detect any strain either in the line between the two poles or in a direction at right angles to this.

Besides the chemical effects noticed in the voltaic cell, there are other effects which are different from those noticed on the outer circuit. For instance, the electrical resistance through the liquid of the cell diminishes when the cell is heated, whereas the resistance in the metallic circuit increases when the conductor is heated. Moreover, there is no polarization effect observable in a conductor carrying a current—that is, no current can be obtained from two portions of a wire through which an electric current has passed. Yet the same portions, on being placed in a liquid capable of being broken up by the electrical current, and after being traversed by a current, exhibit what is termed polarization; they give an electrical current which is opposed to the original current. In discussing these two points of marked difference in behavior between metallic conductors and electrolytes, or liquids which conduct and are decomposed, Prof. J. J. Thomson remarks that these differences are not so marked as to constitute real differences. He points out from the molecular theory that if any polarization took place in the metallic conductor it would disappear so quickly that it would elude observation. In the case of alloys, Prof. Austen Roberts was not able to detect any of the constituents of the alloy at either electrode. Prof.

Thomson again shows that the effect, if any, would disappear with great rapidity. With regard to the difference between conduction in metals and in electrolytes, he states that he has discovered that in an amalgam containing about 30 per cent. of zinc and 70 of mercury the conduction is greater at 80° C. than at 15° C., and remarks that we must remember that the rate of increase of conductivity with temperature for electrolytes diminishes as the concentration increases, and that therefore no sharp line of demarcation can be drawn between the two classes of conductors on this account.

In 1856 W. Weber and R. Kohlrausch endeavoured to obtain a mechanical measure for the strength of an electrical current, and to this end they studied the electrolytic decomposition of water, and stated their results as follows: "If all the particles of hydrogen in one milligramme of water contained in a column of the length of one millimeter were attached to a string, the particles of oxygen being attached to another string, each string would have to be under a tension, in a direction opposite to that of the other, of 2,956 hundredweight (147,830 kilogrammes), in order to effect a decomposition of the water with a velocity of one milligramme per second.*

In the voltaic arc we meet with phenomena which are analogous to the actions which take place in the voltaic cell. There is a difference of molecular activity at the surface of the two carbons. The positive pole becomes much hotter than the negative pole and burns away twice as fast. The transformation of electricity into heat is utilized in a simple process of electric welding, which depends on the phenomenon of the formation

---

* Stallo, Modern Physics, p. 307; Pogg. Ann., vol. xcix, p. 24.

of the voltaic arc. The positive carbon connected with the positive pole of a large battery or a dynamo is immersed in a suitable conducting liquid, while the wire from the negative pole is connected with the rod to which another rod is to be welded. The instant the rod thus connected with the negative pole touches the surface of the liquid intense heat is developed. This heat throws the liquid into the spheroidal state, and increases the resistance between the liquid surface and that of the metal, and raises the latter at the point where it touches the liquid to a white heat. The rod to be welded to the rod constituting the negative pole is then forced into contact with the latter. The rod constituting the negative pole can be held in the hand at one extremity while the other is at a white heat, so quickly is the temperature raised at the end which touches the liquid. The great heat of the voltaic arc is utilized also in the electric furnace which is employed in the production of aluminum, and in various other processes in the arts. This furnace consists of a positive and negative electrode of carbon in the interior of a suitable non-conducting crucible. The voltaic arc is produced between the carbons in the medium which is to be reduced. Another form of this furnace consists of the two electrodes of carbon immersed in a mixture of carbon and insulating earth. In this mixture is placed a crucible containing the substance to be melted; the furnace is surrounded by a layer of non-conducting material.

The street lamp between two pieces of carbon is an electrical furnace without the insulating covering. We generally think of the current flowing from the positive pole to the negative pole. If we use an alternating current both carbons burn at the same rate. Have we not, therefore, in this phenomenon direct evidence that there is an actual flow toward the negative pole? Let us examine more closely the formation of the voltaic arc. In order to produce it, we must bring the positive and negative carbon into contact and then slowly separate them. The resistance of the thin layer of air between the electrodes is broken down by a minute spark due to the difference of electro-motive force of the carbons (about 50 volts), and through the heated air the voltaic arc is established. Numerous experiments show that the light of the arc is due to the incandescence of particles of carbon thrown off from the poles. The heated air is practically dark.

In our discussion of the phenomena of the electric current we have hitherto fixed our minds upon the wire, the liquids through which the current appears to flow, and the earth which forms the return circuit. Our only consideration of the air has been in the phenomenon of the voltaic arc and in the spark discharge, such as is witnessed in lightning. Moreover, we have regarded the coating of a wire as an inert thing having nothing to do with what we have regarded as a flow of something. We have separated all bodies into two classes—conductors and insulators. Our telegraph lines are strung on glass insulators. The wires conducting electricity into our houses are insulated carefully so that the current may not leak away. It would be a strange thought, therefore, to most of us to think of the air, the glass, and the covering of the wire as better conductors of electrical energy than the copper conductor. Yet the tendency of modern investigation is to believe in this hypothesis.

According to this hypothesis, due to Poynting, the electrical energy from the battery or the dynamo spreads out into space and decays into heat when it encounters the wire connecting the dynamo at the central station with the motor in the electric car. It may be said to converge upon the motor in the car, and is there reconverted into motion. The flow of current, or the rate of electrification along the wire, is not in the same direction as the flow of electrical energy in the space not filled with the wire. We see here that the water analogies do not aid us, for the water pumped in at one end of a pipe transmits along the interior of the pipe the energy that is exerted at the transmitting end. None of the energy of the transmitter which in this case is the pump is transmitted through the space outside the pipe.

CHAPTER VII.

THE GALVANOMETER.

Our general survey of the phenomena of the electric current presents the great main features of the subject of electricity as it was known to the world in 1830, if we except the phenomenon of the spreading out of the electric current in the earth and the use of the earth as a return circuit. The knowledge of the action of a battery in the early part of this century strongly resembles the state of the world's knowledge of the action of the human heart in 1630. Two hundred years previous to 1830 Harvey had shown that the arterial blood flowed out from the heart, and, being converted into venous blood, flowed back again. The heart of men and animals was, like the battery, the mysterious source of a strange circulation which was followed only imperfectly by Harvey, for he needed finer and more subtle means of tracing the entire extent of this circulation through the minute vessels which are called capillaries. Of one thing, at least, he was certain: the source of the circulation was the action of the heart, and in the dedication of his book on the circulation of the blood to Charles I he reminds this illustrious prince that, "as the heart of animals is the foundation of their life, the source of everything within them, the sun of their microcosm, that upon which all growth depends, from whom all power proceeds, the king in like manner is the foundation of his kingdom, the sun of the world around him, the heart of his republic, the fountain whence all power, all grace doth flow."

It was not until 1690 that Leuwenhoek showed by means of the microscope that "the blood passes from the arteries into the veins by a network of minute vessels, the thin walls of which allow the fluid plasma to transude into the tissues so as to serve for their nutrition."

When Harvey explained his theory of the circulation of the blood to Charles I, the king must have regarded the heart much as the educated man to-day, ignorant of electricity, regards the power house from which the electrical current apparently proceeds and returns through the bosom of the earth, he knows not how. Indeed, Harvey, although he was certain of the main facts, was at a loss to explain how the blood got from the arteries to the veins. A delicate instrument was needed to help the human eye, and in the hands of Leuwenhoek the microscope proved to be this instrument. This little instrument has greatly extended our knowledge of the immense activities which result from the action of the heart. When, too, Volta was summoued to Paris to explain the electric battery to the great Napoleon, the emperor must have regarded the voltaic cell as a heart, from which a mysterious flow proceeded, no one knew why or how. Volta himself probably never suspected that pulsations in the action of the cell could be detected by an electrical microscope in all neighbouring masses of metal, and indeed in space itself. This electrical microscope is termed a galvanometer, and it plays a part in electrical science very similar to that enacted by the microscope in medical science. The microscope and the galvanometer illustrate what immense advances in our knowledge can be made by suitable instruments. By means of the galvanometer, Michael Faraday and Joseph Henry discovered the principle of the dynamo machine.

It is a strange reflection that these philosophers could have made their discoveries by merely employing the microscope, using the agency of light to discover the manifestations of electricity. We have said that a compass instantly points to a wire through which a current of electricity is passing, and that fine magnetic filings cling to the wire. By distributing iron dust on the stage of a microscope near a wire through which a current of electricity is made to pulsate, an observer looking at the particles of dust through the microscope will see them pulsate also. Furthermore, by winding two little bobbins with fine insulated wire, or, in other words, making little wire spools, similar to small spools of thread, connecting the two ends of the wire on one spool to the ends of the wire on the other spool, placing one spool on the stage of the microscope with iron filings on a piece of paper laid on its end, and placing the other spool on a third spool made of coarse wire the ends of which are connected with a battery, one can observe that when the battery current is suddenly made or suddenly broken the little particles of iron vibrate, showing the existence of currents of electricity in the circuit of wire on the two connected spools. This is the great discovery of currents of induction, made by Henry and Faraday, and it is the foundation of the action of the dynamo and of the telephone. It shows that any change in an electric current on a wire, any pulsation, causes instantly a similar pulsation in any neighbouring wire not connected with the first wire and placed parallel to it.

One of the principal objections to using the microscope in this manner resides in the friction between the fine particles of iron which tends to prevent their free motion. A better way is to suspend a very fine cambric needle from its middle by a spider thread or a bit of coccoon fibre drawn from a white silk thread. Make the needle a magnet by allowing it to rest for a moment across the poles of a strong horseshoe magnet; place the little bobbin of wire not on its end, but on its side, on the stage of the microscope, and bring one end of the needle near the end of the bobbin and focus the microscope upon this end. The needle should be protected from currents of air. It will be a compass pointing north and south and the ends of the bobbin should lie east and west. This microscopic galvanometer can be made to detect very feeble currents of electricity, and will show that any change of strength of an electric current in the bobbin connected with the battery will manifest itself in a neighbouring bobbin in a circuit of wire totally disconnected and independent of the battery circuit. Furthermore, the extent of movement of the needle over a suitably graduated scale of fine divisions will measure the electrical strength of the impulses given to the needle. The microscope thus magnifies the motion of the magnet. Henry and Faraday, however, were ignorant of the use of a microscope for the detection of the very feeble currents which afterward were exalted into the tremendous currents which drive our electric cars and light our cities, and used the unassisted eye to observe the motion of a fine magnetized cambric needle, one pole of which was opposite the end of a bobbin or spool covered with many turns of very fine insulated wire. The longer the needle in general, the larger the deflections. In order to avoid weight in the suspended magnet it was made of a bit of magnetized steel, which was provided with very light hairlike pointers. The ends of these pointers moved over a graduated scale. To-day, instead of pointers, a long beam of light is used as an index. It has no weight, and it takes the place of pointers at least three feet long. To use a beam of light the suspended magnet is stuck upon a tiny mirror of very thin silvered glass and a beam of light is reflected from this mirror upon a distant scale or into a suitable telescope. A small movement of the magnet thus results in a large movement of the beam of light on the distant scale. This is a method of magnifying small movements which is much used in physics, and it has even a practical application, for it showed the possibility of telegraphing beneath the ocean by means of the Atlantic cable, and it has been constantly in commercial use on the cable up to this date. The device of a tiny mirror on a suspended magnet, applied by Lord Kelvin, has been of great practical use in submarine telegraphy. The microscope made possible the investigations which have led to the germ theory of disease, the antiseptic treatment in surgery, and founded the subject of physiological botany. The galvanometer has shown that electrical actions pervade all matter, and that there are electric waves in the ether of space.

With a delicate galvanometer the field of our knowledge of electrical activity is enormously increased. We immediately discover that the slightest movement of a magnet near a wire is accompanied by an electrical disturbance on the wire. If we thrust one pole of a magnet into a spool covered with wire the ends of which are connected with the terminals of wire wound on a similar spool, a little magnet provided with a mirror, and suspended opposite the end of the latter spool, will show a momentary electric current in the electric circuit constituted by the two connected bobbins. It dies out immediately, but it is instantly renewed when we pull out the pole of the magnet instead of thrusting it into the bobbin. When we pull the pole out, the suspended magnet moves in one way, and when we thrust it in, it moves in an opposite way. Furthermore, if we thrust the south pole of a magnet into the bobbin the galvanometer needle moves in one direction, and when we thrust in the north pole it moves in another direction. Since we perceive that any movement of a magnet near a coil of wire is attended by an electrical disturbance in the wire, it is natural to suppose that we can produce electrical disturbances by keeping the magnet fixed in position and by moving the coil of wire. Experiment will speedily verify this conclusion. We find that a movement of our little bobbin or coil of wire in the air near a pole of a magnet excites currents in the circuit connected with the bobbin.

Since the earth appears to be a magnet, we should expect, therefore, to find that if we move our little coil in the air we should obtain an electrical current. Experiment shows this to be true, and it still further illustrates what the delicate galvanometer revealed to Faraday and Henry. There was something in the space outside a magnet which could be made evident by moving either a wire or the magnet. The slightest change in the current passing through one wire makes itself manifest across space in a distant wire. The galvanometer can detect slight molecular disturbance; it can also reveal mysterious effects in the ether of space. One sometimes smiles at the microscopist who disputes with a fellow-microscopist on the possibility of measuring spaces of one hundred thousandth of an inch. Yet one should reflect that the antiseptic treatment in surgery, which saves thousands of lives every year, is due to the perfection of the microscope. To those unlearned in electrical science, too, the tiny movement of a needle or the excursion of a spot of light over a scale seems hardly worthy the attention of a liberally educated man. Yet the modern dynamo and the telephone owe their existence to the study by Faraday and Henry of these minute movements.

I have endeavoured to explain the construction of the galvanometer, which I have termed the electrical microscope, and in the following chapter I shall endeavour by its aid to explain the action of the dynamo machine and the electrical motor which is now used to propel electric cars. The galvanometer, we have seen, consists in its essentials of a coil or bobbin of wire like a spool of thread with a tiny magnet hung by a spider thread near one end of the spool. When a current of electricity flows through the spool it makes the spool an electro-magnet with two poles, and the pole near one end of the spool consequently attracts one pole of the little suspended magnet and repels the other, thus causing the tiny magnet to turn around its axis of suspension. When the current ceases the suspended magnet returns to its original position, pointing north and south. The amount of its turn measures the strength of the electro-magnet.

There is another instrument called the electrometer, which is used by electricians to measure the electromotive force instead of the current. The galvanometer indicates the current which results from a certain electro-motive force; it does not measure directly the electromotive force which gives rise to the current. For instance, it will measure the current given by a silver spoon and an iron spoon immersed in a tumbler of water, but it does not tell us directly what is the electro-motive force between the iron and the silver. This electromotive force is similar to the pressure of the water or the steam at the power house, under which pressure the circulation is caused in the pipes issuing from the power house. It has been found that any two metals immersed in a liquid which acts unequally upon them give rise to an electro-motive force, which varies in strength with the character of the metals. It is therefore useful to have an instrument which will serve as a sort of gauge of electrical pressure. This instrument is called a voltmeter, the volt being the unit of electro-motive force. The indications of this instrument are graduated by means of what is called the Latimer Clark cell. This is a voltaic cell which produces a constant electro-motive force and which therefore serves as a unit.

CHAPTER VIII.

THE DYNAMO MACHINE.

We have remarked that the ancients could not have possessed dynamo machines or telephones, for they were ignorant of the art of covering wire by cotton or silk, and it is doubtful whether they knew the art of drawing wire. If by any cataclysm or widespread catastrophe the European nations should disappear from the face of the earth and some East Indian tribe ignorant of electricity should alone survive, and through the slow ages rise and possess the American continent, their archæologists might find the ancient ruins filled with copper wires. There is no trace of such wires in the ruins of Egypt or of Greece. In reading the Life of Faraday, and his account of his discovery of the principle underlying the great advances in electricity, we wonder why the phenomena which he discovered had not been discovered before, and by men with less mental equipment. When we consider, however, the form of galvanometer with which he worked, and the limited supply of insulated copper wire of different degrees of fineness which was at his disposal, our wonder disappears. With a modern galvanometer a senior in Harvard University of fair intelligence could not fail to discover the laws of magnetic induction, for an accidental movement of a magnet near a coil of wire would certainly reveal these laws. Faraday's merit consisted not so much in the discovery of the phenomena as in his mental conception of lines of force pervading all space.

It will be interesting, therefore, to compare Faraday's instruments with the more sensitive modern apparatus. In the first place, in common with Tyndall, he used a form of needle galvanometer, and he observed the movements of the ends of the needles over a graduated circle just as we observe in a pocket compass the movement of the needle of the compass over the divisions of the circle placed beneath the end of the needle. It is evident that a small angular movement of the needle can not be very much magnified on a circle of small dimensions. Faraday's circle was not more than four inches in diameter. In the modern galvanometers what would correspond to circles of ten feet in diameter are often used, thirty times the diameter of Faraday's scale. Let us examine further Faraday's galvanometer and the modern form, remembering that we are now dealing with an electrical microscope which has revealed the great world of electrical activity in which we move.

Faraday's and Henry's galvanometer consisted, in the first place, of two ordinary needles—cambric needles—which were magnetized and fixed one above the other (Fig. 8) to a rigid bar, A B. The poles of one needle were opposed to those of the other—that is, a north pole was placed above a south pole, or a south pole above a north pole. The object of this arrangement was to weaken the effect of the north pole of the earth, so that the arrangement of needles should not be held so strongly to the north and south direction, and should yield to a very slight attracting force at right angles to the direction of the earth's pull on the compass. The two needles were suspended to a fixed support, C, by means of a strand of cocoon fibre. This arrangement will show the presence of an ordinary bar magnet five or six feet away. If only one needle is used, it will not show the attracting force of the same bar magnet if the latter is brought within three feet of it, so strongly would one needle be held by the north and south poles of the earth.

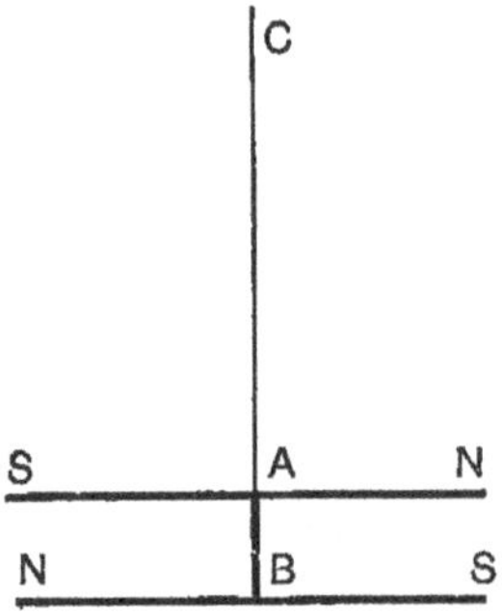

Fig. 8.

Fig. 9.

Beside Faraday's little suspended magnet I have represented in Fig. 9 the suspended magnets in a modern galvanometer. At the end of a light aluminium or rigid glass fibre are placed collections of tiny magnets. The north poles of those at A are opposed to the south poles of those at B, and the south poles at A are opposite the north poles at B. A little mirror of very thin silvered glass is placed at N. The weight of Faraday's suspended apparatus was at least 400 milligrammes, while the weight of the modern form is only 80 milligrammes—hardly more than that of a butterfly's wing. Moreover, the later form is suspended from the fixed support by a quartz fibre finer than the finest hair. This quartz fibre is very strong and possesses hardly any torsion—that is, any power to resist a twist. The suspended magnets will therefore turn about the axis of their suspension under the influence of the slightest attracting force. The use of this quartz fibre, due to Prof. C. V. Boys, constitutes a great improvement in the suspended apparatus of the galvanometer. It is obtained in the following manner: A bit of quartz is suitably attached to the arrow of a crossbow. It is then melted in an oxyhydrogen blowpipe, which fuses it into a globule of quartz glass. At the proper point of fusion the crossbow is discharged and the arrow flies away, carrying a fine thread of the melted quartz. Threads of different degrees of fineness can be obtained by regulating the strength of the crossbow.

The suspended apparatus, therefore, of the modern galvanometer is at least five times lighter than that used by Faraday. It will turn about its axis with far greater readiness under the influence of slight attraction, such as is exerted by a distant magnet. But the greatest improvement consists in magnifying its movement around its axis of suspension by means of the reflection of a spot of light from the tiny mirror. By means of such a reflection we can practically increase Faraday's graduated circle of four inches in diameter to a circle of one hundred and twenty inches in diameter; and, moreover, we use as a long index a beam of light, without weight. A suspended apparatus of this form will show the presence of a bar magnet at a distance of at least thirty feet. In the Jefferson Physical Laboratory it responds to the passage of the electric cars four hundred feet away.

In both the earlier and later forms of galvanometer the lower suspended magnets are hung at the centre of little coils of insulated wire of many turns. When an electric current flows through the wire of the coils it makes the coils an electro-magnet, and the suspended magnets tend to place themselves in the axis of the coils—that is, their poles are attracted to the poles of the electro-magnet. The attraction of the earth tends to turn the magnets back into the north and south line, or into the magnetic meridian, and they therefore take up an intermediate position, and this position serves as a measure of the force of the electric current in the coils.

A lady's workbasket, with the addition of a toy magnet and with fine covered wire instead of thread, affords all the material for illustrating the great principles of magneto-induction, which underlie the action of the dynamo machine and the telephone. Two cambric needles can be magnetized by placing them for a moment upon the poles of the toy horseshoe magnet so that they constitute its armature. They can then be stuck, after the manner of the needles in Faraday's galvanometer, through a bit of light wood; and the two needles can be suspended by the ultimate fibre drawn from a white silk thread. A spool wound with fine silk-covered wire instead of thread can be placed with one end opposite one pole of one of the needles, and we shall have then a model of a Faraday galvanometer in all essential respects. The pole of the cambric needle will turn toward the end of the spool of wire when a sufficiently powerful current circulates through this spool, and by its movement will thus indicate the strength of this current.

Suppose that in the adjoining Fig. 10, N S represent the lower suspended cambric needle as it appears when we look directly down the suspending fibre, and that S represents the spool covered with wire. Let another similar spool, S′, be placed at some distance, and let the ends of its wire, B, D, be connected with the ends of the wire, A, C, of the first spool. If we should now thrust the north pole of a thick bundle of magnetized needles into the coil S′ the little needle N S will move in one direction, and when the magnet, N

S, is pulled out of the coil it will move in a reverse direction. If, therefore, before we pull out the magnet we also reverse the connections of the wires—that is, connect D to A and C to B or the reverse—the needle will always move in the same direction whether we thrust the pole N into the spool S′ or pull it out. The person who thus assists the experiment by changing the connections of the ends of the wires can be called a commutator. He commutes the direction of the cur-

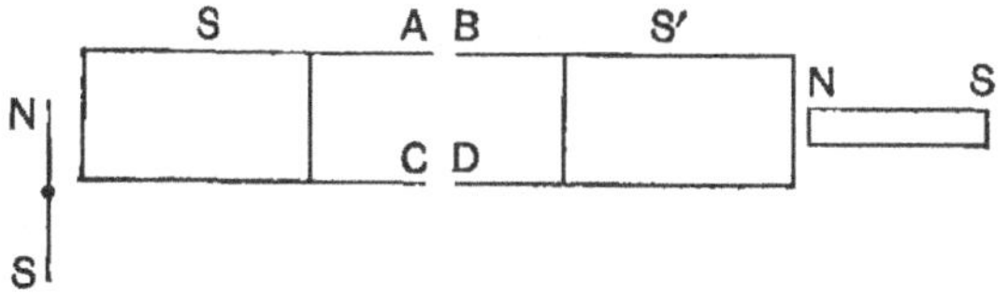

Fig. 10.

rents which fly in opposite directions through the spool S′, so that they shall always pass through the spool S in the same direction, and so that we shall thus have—if he commutes swiftly enough—a steady current through S instead of a to-and-fro or alternating current. It is evident that it would require very little ingenuity on the part of the person who acts as commutator to arrange a mechanical apparatus which would relieve him of his task and would commute automatically—that is, place the ends A, B, and C, D, in communication when the pole N is thrust into S′, and the ends A, D, and C, B, when it is withdrawn. There is an interesting analogy between the action of the automatic commutator and the eccentric of the steam engine, which admits steam first at one end of the piston and then at the other. In the early forms of the steam engine which were employed in the mining districts in England steam was admitted beneath a piston by turning a cock by hand. The steam,

8

having lifted a load by means of the piston, was discharged from the cylinder so that it might move back ready again for an upward stroke. It is said that the boy who had charge of admitting the steam properly devised the eccentric to lessen his labours—to do the work for him while he occupied his leisure in playing marbles. This story may be a fable, but it serves our purpose well for an illustration. The eccentric, it is known, is a mechanism connected with the moving parts of the piston rod, which reverses the admission of steam from one side of the piston to the other, in order to secure a continuous movement to and fro of the piston rod. The eccentric makes the modern steam engine a practical continuously moving machine, and the electric commutator has made possible the modern dynamo and electric motor. I dwell at some length, therefore, upon this simple practical detail, to illustrate how greatly the advance of science depends upon the advance of the mechanic arts. The machines for covering copper wire and the perfection of the galvanometer and commutator have made possible the rapid practical employment of the feeble currents discovered by Faraday. The commutator has exalted them from currents which were barely perceptible to Faraday to currents capable of driving electric cars weighing many tons.

I have employed the following simple apparatus to illustrate the action of the dynamo and its commutator, which requires very little mechanical skill to construct. A disk of wood about a foot in diameter (Fig. 11A) has a pointed stick driven through its centre, thus making a large Japanese top. The disk being very near one end of the rod which passes through its centre, such a top will continue to revolve in the same spot on a smooth floor for a long time by merely twirling the rod between the fingers. Stick now a little flat spool of insulated copper wires (S′) upon the wooden disk, lead its wires up the rod, and bind over their bare ends tightly two little shieldlike pieces of bright sheet copper or brass, so that these shields partly embrace the rod, being insulated from each other by the wood.

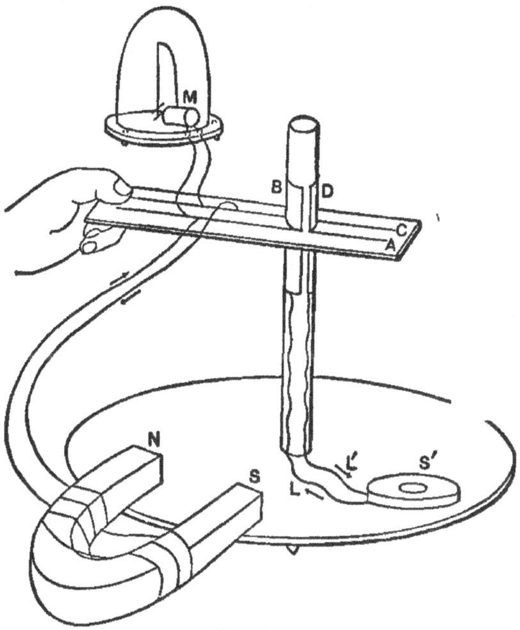

FIG. 11A.

These shields form the ends, B, D, of the moving spool, S′. On a piece of wood in which a hole is bored to admit of the top turning run the wires A, C, to the coil M of the galvanometer, and remove their covering, so that the bare wire shall touch the ends B and D. When the top revolves, the coil S′ passes under the south pole S of a horseshoe magnet, and an electric current flies through the spool S′ in the direction of the arrows L L′. When S′ passes under the pole N a reverse current is induced in it in the direction of the arrows. At the same time, however, the ends of the wires B and D are reversed, so that D touches A and B touches C, and the current is still delivered to M in the same direction.

This model represents the modern dynamo machine in certain of its essential features. It shows that currents of electricity are excited in a coil of wire passing rapidly near the poles of a magnet; that a current is excited in the coil in one direction by movement near a south pole, and in the opposite direction by movement near a north pole; and that a commutator can be made to direct these opposite currents in the same direction through another coil. The next great advance in the dynamo machine was in a simple process of strengthening the poles of the horseshoe magnet by the current which it excites in the moving coil, S′. This was done by running the current, which we have shown can be tested by the galvanometer, M, around the poles of the horseshoe magnet, N and S, wrapping the wire around the pole N in one direction and around the pole S in the opposite direction, so as to excite opposite poles; for direct experiment shows that the direction of the winding about a piece of iron or steel determines whether a north or a south pole is made. By means of this arrangement of the circuit a feeble current in the spool S′, strengthens the poles N and S. This strengthening in turn leads to a stronger current in the coil S′, and this stronger current again strengthens still further the poles N and S. This process goes on like the law of compound interest, until the work done in twirling the top is just equal to the work done in overcoming the repulsions and attractions between the poles of the little moving coil S′ and the poles of the electro-magnet N S. That there exists a repulsion instead of an attraction between the moving coil S′ as it approaches the poles N and S is evident from this consideration. Suppose that an attraction should exist. After the top is once started it would go on forever, for the coil S′ would be attracted to the south pole, S, and then to the north pole, N, and so on. Our top would then be a perpetual-motion machine, the essence of perpetual-motion machines being in their ability to move without consuming any work.

If we should now turn our top by a steam engine instead of by the fingers, we should have the entire arrangements of a modern dynamo machine. Instead of one coil, S′, we should have a number, and the poles N and S would be more advantageously placed. Still, our model represents in its essential features the modern dynamo, and also the electric motor. Fig. 11B shows the practical arrangement of what is called a series dynamo, the coils on an iron ring being connected to segments of the commutator, X, while the current is taken off by the brushes, *b b′*, and is made to flow around the field magnets, N and S, so as to make north and south poles, as the figure indicates. It is well to note here that a steam

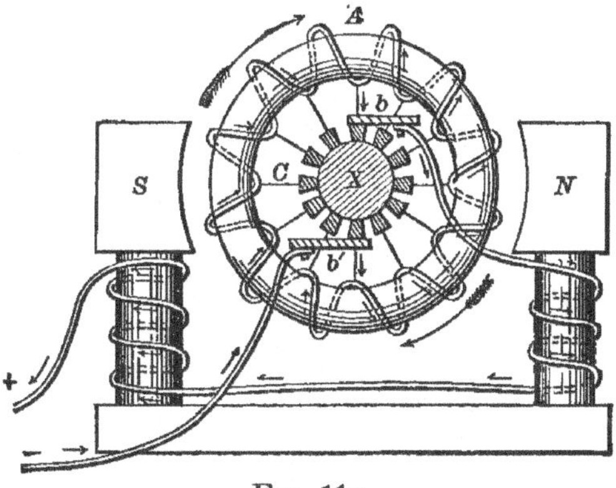

FIG. 11B.

engine is employed to drive all dynamos, and that behind all our wonderful uses of electricity is steam. We see, therefore, that electricity can not supersede steam as a source of power until some method of obtaining electricity direct from coal is discovered. At present there is apparently no feasible method. The popular impression, therefore, that electricity will speedily supersede steam has no foundation in fact. This impression may be prophetic, but steam is now the source of all our electricity in the transmission of power, if we except water power.

CHAPTER IX.

SOURCES OF ELECTRIC POWER.

The practical applications of electricity afford rich illustrations of the transformations of energy. The coal which generates the steam which drives the engine was produced by electro-magnetic waves from the sun. In the dim past these waves, in the form of light and heat waves, nourished the great ferns and palm trees, and all that luxuriant vegetation which now constitutes our chief reliance for power. Through the combustion of the coal we are enabled to drive the dynamo which produces the electric energy, which is again transformed into motion.

Hardly a tenth of the electro-magnetic energy stored up in the coal is given out by the steam engine. Our principal source of electricity to-day is the steam engine; yet it can be maintained that electricity is back of the steam engine. It is in the coal. It produced this coal, and could make itself manifest to a large degree if we knew how to transform it. Many attempts have been made to obtain electrical power direct from coal, but they have not been successful. I have already referred to thermo-electricity. We have seen that it is possible to construct a furnace with junctions of suitable metals imbedded in its fire pot so that the combustion of coal will produce electrical currents in these junctions. Indeed, an electric arc light has been formed by such means. The furnace, however, must be cumbrous, and the junctions expand and contract under the changes of temperature and are soon ruined.

We have seen that the modern dynamo machine is a comparatively simple apparatus, merely a number of coils on a revolving shaft surrounded by fixed pieces of iron, around which the currents formed in the revolving coils are made to circulate. The working of the dynamo machine is entirely dependent upon the steam engine which drives it. In the best dynamo machine only about 15 per cent of the energy supplied by the steam engine is lost. When we consider the friction of the bearings and the resistance encountered, the economy of the transformation of energy from that of steam to that of electricity is very perfect. The modern dynamo seems to have reached its highest development.

The spectacle of the transformation of energy by the dynamo in our great cities is most impressive. At the central station are immense steam engines which are whirling the movable coils of the dynamos on axles which run at about one thousand revolutions a minute. The earlier dynamos could almost be lifted by one man. The dynamos of the central stations weigh tons. Thus the mechanical engineer, taking the principles discovered by Faraday, has adapted it to a very perfect machine, and has made one of the greatest transformations of energy witnessed in the mechanical arts.

Another transformation of energy resulted from the construction of the dynamo machine. We have shown that the current produced by one dynamo can, if led to the wires of a second dynamo, make the movable coils of the latter revolve. Thus the second dynamo becomes a motor, and can be used to turn shafting or set in motion any form of machinery. Every electric car has a dynamo motor connected with its axles, and the current produced by the great dynamo at the central station sent over the trolley wire propels the electric cars. Now, electro-magnetic engines were not unknown in the time of Faraday. They were, however, mere lecture-table models, and were run by the current obtained from batteries. The lecturer of twenty years ago often took such models as a text to show the impossibility of obtaining power in this way, for a short calculation of the amount of zinc consumed in the battery compared with the power produced showed the want of economy in such electric motors. Moreover, the mechanical construction of the early electro-motors was very defective. The idea of reciprocating motion, like the piston of a steam engine, was the ruling one. So machines were made with iron cores suspended from a species of walking beam, like that of a steamboat. These cores were alternately sucked into or repelled from coils of wire through which currents of electricity were circulating. The early investigators were appalled by the quickness with which the magnetic forces of attraction and repulsion decreased with the distance. These forces are proportional to the product of the attracting masses and inversely as the square of the distance. Thus the force at the distance of one inch is only one fourth of that at one half an inch. The play of the reciprocating magnetic engine was therefore very small, and the cost of the zinc which was consumed in producing this small play was great. There seemed, therefore, little hope in the employment of magnetic motors. With the discovery of magnetic induction and with clearer ideas of the magnetic field surrounding magnets and electro-magnets hopes in this form of motor revived and were realized.

Currents of electricity could be produced of almost unlimited strength by the dynamo. Very powerful attracting magnets could therefore be made. The revolving coils of the dynamo cut the lines of force of the stationary electro-magnets very close to the face of the poles of these magnets, where the number of lines of magnetic force are greatest. The distance between the revolving coils and these poles in some cases is less than one eighth of an inch. When, therefore, an electric current is sent through a dynamo, the force of attraction between the revolving coils and the stationary coils is very great, since the distance between them is so small; and the commutator changes, as I have said, the poles, so that a continuous rotation of the movable coils is produced.

The mechanical arrangement which we have called the commutator is an important factor in the production of a magnetic motor. A clearer conception, however, of the way magnetic lines of force spread out from magnetic poles has led to the perfection of the magnetic motor. Every one is familiar with the way iron filings arrange themselves near the pole of a magnet. They form radiating lines which spread out from the poles as centres of disturbance, and seem to arch from the negative to the positive pole, forming what may be termed magnetic circuits between the poles. If our magnet is made in the form of a horseshoe, the arching is more pronounced; the filings crowd into the space between the two poles, and fewer extend into space. By making the horseshoe nearer and nearer the form of a ring, and bringing the north pole almost into contact with the south pole, the air gap between them being very small, hardly any lines of magnetic force extend into the space about the poles. They are concentrated in the air gap. They form almost a closed magnetic circuit through the ring, and the magnetic field in the air gap can thus be made very intense by the number of lines of magnetic force which crowd into this space. In the modern dynamo and the electric motor the stationary electro-magnets are made as near the ring form as possible. The revolving coils are placed in the air gap, where the magnetic lines crowd from one pole of the stationary electro-magnet to the other. Very few lines of force are suffered to stray out of the air gap into outer space. They are nearly all cut by the revolving coils. In considering this form of construction, we readily see how imperfect the early electric motors were. Very few of the lines of magnetic force were utilized in attraction. Most of them strayed into the air and were lost, so far as useful work was concerned.

With the perfection of the electric motor the steam engine became a more important factor in civilization than ever. Electricity appeared to be ready to usurp its place. It could not be produced, however, economically without steam. The servant could not take the place of the master. A new method of distribution of power, however, has sprung into existence, for the electric current generated by steam can be carried miles from the producing station, and can be transformed again into motion. The present limit of distance to which steady currents of electricity can be economically carried for conversion into power is about five miles, for the electricity tends to escape from the wires to the earth. Moreover, the resistance offered to its flow by the wires becomes too great. I have said steady currents; we shall see later that the distance to which fluctuating currents can be carried is far greater than five miles. In certain manufacturing establishments power is carried from one building to another by belts and shaftings or by means of ropes. The loss in transmission on account of friction and rigidity of ropes is very great if the distance exceeds a few hundred feet. It is estimated that in transmitting power by means of electricity five miles at least fifty per cent of the available energy is lost. The only ways, however, that power can be transmitted five miles from a central station are by means of a current of electricity led on wires; by means of storage batteries; by cables; or by compressed air. In all of these methods the steam engine is the source of the energy. The electrical method is the most flexible one, for wires can be easily led in almost any direction. In the great cities at present the electrical energy which is used to light the city at night is employed in the daytime in running the machinery of hundreds of workshops.

In the many examples which we thus see of the transformations of energy accomplished by the invention of the dynamo we find the great truth of the conservation of energy constantly exemplified. The electrical engineer calculates the efficiency of his dynamo by measuring the amount of work the steam engine does in turning the swiftly revolving coil of the dynamo and comparing this with the amount of work the electricity can do. He assumes that if the mechanism of conversion were perfect the energy produced must be equal to that expended. Tyndall, in his Heat as a Mode of Motion, showed that friction could be converted into heat by rapidly revolving a closed tube containing water between a suitable clamp. By means of dynamos and motors I have modified this experiment in the following manner: Let a steam engine set in action a dynamo; then the dynamo an electric motor. On the continuation of the shaft of the latter place a hollow tube partially filled with water and tightly corked. If now a friction brake clasps this tube and the motor is set in revolution, in a few seconds the water boils, steam is formed, and the cork is thrown across the room by the expanding steam. Thus the steam, through various transformations, produces again steam, and the work done against the friction brake can be estimated by the amount of water raised to 212° F. This in modern form is the celebrated experiment of Count Rumford by means of which he set the world to thinking about the conversion of work into heat.

We have followed the work of steam from the engine to the dynamo, where its energy is converted into electrical energy, from the reconversion of this energy by means of the electric motor to motion, and we have seen that steam is still the great moving power in the machinery of the world, while electricity can be said to be only its servant, nimble and pliable. We have said nothing of the transformation of water power into electricity. Why is not this source of power cheaper than steam to produce electricity? The modern stationary engine has been brought to such a state of perfection that a horse power can be obtained from a pound and two tenths of coal, while the cost of regulating a water supply costs more than this. At present electricity can be manufactured in Buffalo from coal cheaper than it can be transmitted from Niagara Falls to Buffalo, a distance of about thirty miles. We shall return to the subject of the electrical transmission of power in a subsequent chapter. Now we are considering the transformation of energy by means of steady currents of electricity. Long-distance transmission of power can not be accomplished by steady currents unless some efficient form of storage battery should be invented. The storage battery illustrates another transformation of energy which awakened great hopes, but which has not yet fulfilled the expectations of mankind. The storage battery or accumulator in its commonest form, we have seen, consists, before it is charged, of red oxide of lead with electrodes of lead. The oxide and the electrodes are suitably immersed in dilute sulphuric acid, and a strong current of electricity is sent from one electrode to the other. The oxygen resulting from the electrolysis of the water converts the red oxide of lead at the future positive pole of the battery into peroxide of lead and into metallic lead at the future negative pole of the cell. When the charging current is removed and the peroxide-of-lead pole is connected with the porous metallic lead pole, a current of electricity is produced and the peroxide goes back to a lower oxide. The transformation from the energy of steam to that of electricity in this case is by means of chemical action, and chemical action has not yet produced electricity economically.

I suppose that no subject in the development of electricity has received so much attention as that of storage batteries. It was quickly shown that the electro-magnetic engines run by batteries were far more expensive to run than the steam engine; for in order to smelt the zinc for the plates of a battery it was necessary to expend about sixty times its own weight in coal, while a pound of coal could produce a much greater amount of horse power than a pound of zinc. If we should supplant the steam engine at a central station by a great battery plant we should consume zinc in sulphuric acid instead of coal under a boiler. The consumption of zinc in sulphuric acid generates energy in the form of heat. By connecting the poles of a battery with a copper wire we convert this energy into electricity, and reconvert it into heat wherever the resistance of the line is sufficient. The transformation of energy thus proceeds from chemical action through electrical action back to heat. Batteries in which chemical action is used to consume zinc are called primary batteries, and inventors are still busy in endeavouring to produce an economical chemical source of electricity, but none has yet been discovered. Indeed, the use of primary batteries is daily growing more limited, and the dynamo machine and storage batteries are taking their place. A storage battery can be made which will give back nearly eighty per cent of the energy which is used to charge it. Its life, however, is limited. It is fully as long, however, as that of the best primary battery.

While the storage battery has not fulfilled the hopes it excited, it has gradually come into commercial use in steadying the load at central power and light stations. In case of an accident to the machinery the current from a storage battery can be turned on the line. It acts like a spare reservoir of water for sudden emergencies.

It is probable that some other substance better fitted than lead for the use of accumulators will be discovered. One can to-day use peroxide of lead and zinc in a storage battery and obtain more powerful currents than with the use of lead for both elements of a storage cell. This is a suggestion of De la Rive, and it has been developed by Prof. Main into a zinc-lead storage battery. The zinc, however, consumes rapidly, and there are chemical reactions which are troublesome and destructive. I have found the following method, however, of transforming energy by means of zinc extremely useful in the laboratory. In order to understand it one must look at an old form of primary battery which even now is used in many laboratories where a portable source of electricity is required. It is called the dip battery, and consists of alternate plates of carbon and zinc, which are immersed, when used, in a solution of bichromate of potassium. The carbon constitutes the positive pole and the zinc the negative. The zinc is consumed by chemical action, and a current of electricity is produced by this energy along the wire connecting the carbon pole with the negative. When the battery is not in use the plates are lifted out of the bichromate of potassium. The chemical action is of short duration, and an hour's use generally depletes the battery. Instead of carbon I have used porous partitions filled with oxide of lead converted into peroxide of lead. In charging, the negative electrode is a simple lead plate. After the cell is charged the negative lead plates are lifted from the sulphuric acid and amalgamated zinc plates are lowered in their place. We have thus a modern dip battery immensely more powerful and serviceable than the old form of primary dip battery. This use of zinc led me to an interesting method of regenerating an old used-up storage cell which was incapable of storing up the energy supplied to it. The use of a zinc electrode restores the positive lead plate to usefulness by breaking up an injurious sulphate of lead which is formed. Two recent observers, MM. Cailletet and Coladeau, have succeeded in storing in the metal palladium under six hundred atmospheres pressure more than ten times the amount of electric energy which can be stored in the same weight of lead oxide.

The method, however, of transforming energy by means of electrolysis does not at present seem to be economical.

There is another method of obtaining electricity from coal without the formation of steam, and this consists in the employment of gas engines. Ordinary illuminating gas mixed with the proper proportion of air is exploded in a cylinder, and the explosion drives a piston to and fro. Certain writers contend that it is cheaper to drive a dynamo by such an engine and produce light by electricity than to burn the gas direct. Great improvements have been made in this method of producing electricity. Gasoline, for instance, can be used instead of illuminating gas. With the gas engine all the transformations of energy which we have shown can be accomplished by steam are possible. We have thus another possible method of the transmission of power. Gas can be distributed from a central station instead of electricity, and can be converted into electricity when the latter is wanted. A gas engine does not require the services of a skilful engineer and of a fireman. Merely lighting the gas torch in the engine with a match sets the engine in motion. The tendency of the times is to use more concentrated fuel than coal. Thus we have oil engines and naphtha launches. Maxim guns may also be considered a species of gas engine. This powerful weapon is capable of discharging six hundred shots a minute. The recoil from the first shot brings another cartridge into position where it is fired, and so on. The gun is self-acting after the first shot. With the powerful explosives at our command the power we could exert in engines is enormous. The difficulty is in properly controlling and employing the energy of the explosion.

The subject of gas engines brings us to a consideration of compressed-air motors. Many believe that the employment of compressed air as a source of power on street railways has not been sufficiently tested. The modern mechanical engineer has succeeded in making cylinders of steel which can stand with safety the enormous pressure of 60,000 pounds to a square inch, and it does not seem impossible that we may yet see compressed-air motors actuating machinery which is now driven by electricity. In the Calumet and Hecla copper mine power is supplied in the mine by compressed air. The loss at the distance of three miles is about fifty per cent, while the loss by electricity is not far from the same amount. Greater safety results from the use of compressed air in mines than from the employment of electricity, for there is no danger from fire by this method of transmission of power.

Maxim, in his interesting experiments on aërial navigation, has entered carefully into an estimate of the power that can be developed by various kinds of motors in comparison with their weight, and gives the following: Hot-air engines, 200 pounds to the horse power; Brayton's oil engine, 75 pounds to the horse power; electric motors fed by secondary batteries, 130 pounds to the horse power; gas engines (Otto), 50 pounds to the horse power; steam engines, with condenser, pumps, and everything complete, 25 to 50 pounds to the horse power.

Maxim's complete steam motor as it existed in 1894 weighs 2,040 pounds, which includes the boiler, engines, gas generator, pumps, and 200 pounds of water in the boiler, but this does not include the supply of fuel, the water in the tank, or the condenser. The highest power developed was 363 horse power, which gives 5·6 pounds to the horse power. An atmospheric condenser can be made weighing no more than one pound to the horse power.

The transformation of enhanced molecular activity through the varied chain of motion of machinery, whirls of magnetic lines, to the storage battery, where the molecular action again reappears, is not more striking than the transformations which can be accomplished by the battery. Thus machinery can be turned, magnetic fields formed. In both transformations incandescent lamps can be lighted and water converted into steam. Still further, the electric light generated by the molecular activity of the steam or the molecular activity of the storage battery can set up or modify the molecular activity of a sensitive photographic plate, producing before accomplishing this result waves in the ether of space. It is probable that we know little of the effect of electro-magnetic waves in the ether in transforming or changing the rates of molecular motion. Attempts are being made to ascertain the effect of electrolysis on plant growth. Certain portions of planted fields in France have been submitted to the effect of the electric current—they have been charged, so to speak, like storage batteries—while other portions have not been thus treated. It is said that a difference in the rate of germination and growth of plants has been detected. The electrical treatment apparently has stimulated the vegetation. Siemens submitted certain plants to the rays of the electric light for a long period of time, and greatly enhanced their growth. The molecular action in one case was stimulated at the root and in the other in the leaves, and the ultimate source of the stimulus was in the coal measures, produced by a luxuriant vegetation of ages ago, which in turn, according to modern belief, had its source of growth in electro-magnetic waves from the sun.

In our account of the transformations of motion into electricity we have dwelt largely on the conversion of electricity again into motion. The production of light by electricity has become a matter of such daily observation that we have ceased to wonder at it. Attempts have also been made to heat economically by electricity. All that is necessary is to increase the resistance of the electrical circuit at the point where we desire to produce the heat.

We have followed the transformations which result from burning coal, the fossilized vegetation of a former age; let us now examine the transformations exerted by man considered as an engine. The food he consumes answers to the fuel we put under the boiler of the steam engine. The source of this food is also vegetation, and the ultimate source of vegetation is, according to modern theories, the electro-magnetic radiations of the sun. Animals and man take in oxygen and give out carbonic acid. It was pointed out by Joule that man more nearly resembles an electro-magnetic engine than a steam engine, and he also showed that man as an engine is far more efficient than any form of engine which he can construct in which there is a consumption of fuel. Man may be roughly compared to a voltaic cell. In the cell we have sulphuric acid and zinc in the shape of food, while the resulting energy in the form of electricity can be transformed into motion or heat. In man food, together with oxidizing processes, produce also motion and animal heat. Vain attempts have been made to measure the efficiency of the man engine by weighing the food consumed and measuring the work done. The transformations of the food, however, are so varied and so subtile that it is difficult to estimate them. In general, however, the amount of work a man does bears a certain proportion to the fuel he puts into his boiler. It is interesting to notice that in the modern practice of medicine the conservation of energy is recognised. Heat is supplied to invalids in order that the human engine may not be compelled to supply this animal heat. Too great an output of heat in the case of fevers is checked by the application of cold water. Various instruments, similar to indicators used to ascertain the horse power of steam engines, are employed by physiologists to study the action of the heart considered as a pumping engine. Perhaps the French physiologist Marey has more than any one else called attention to the importance of studying the action of different parts of the human engine.

Now the plants also constitute forms of engines which use the products rejected by men and animals. By means of the radiant energy of the sun they are enabled to decompose carbonic acid. One can compare a plant to a storage cell in which a current of electricity decomposes the liquid into oxygen and hydrogen, and forms materials which again give electricity and motion and heat. The sun enables the plant engine to work. The rays—so-called chemical rays—that is, the shortest waves of light, are most effective in producing the decomposition of carbonic acid by the leaves of plants. These rays are those which are the most absorbed by the foliage of plants. This is shown by the ordinary photograph of a landscape. The leaves are black except when they reflect light. They have absorbed the rays of light, and have decomposed carbonic acid and water. The efficiency of the plant engine is still more difficult to obtain than that of the man engine, for the subtile transformations of the chlorophyll are many. We have the innumerable coal-tar dyes, which seem to have as many peculiar properties as they have colours. Also in photography we have hydroquinone, rodinal, amidol, and a host of other developers which are used to give pictures of the very plants which by previous transformation have produced them.

CHAPTER X.

TRANSFORMATIONS OF ENERGY.

I HAVE dwelt upon the construction of a galvanometer and of the dynamo machine in order to emphasize the great bearing that properly constructed machines have upon the progress of science. Faraday's galvanometer was not a very sensitive instrument; it was analogous to the one-lens microscope of Leuwenhoek; yet it was sufficient to show the great law of magneto-induction, that the movement of a coil of wire near a magnetic pole produces a current of electricity in the coil, and when it was aided by powerful auxiliaries it showed that any change in an electric current produced a current of electricity through the ether of space in neighbouring conductors. These powerful helps consisted of a strong voltaic battery of large plates, and a great many of them, and of a powerful electro-magnet which Joseph Henry had shown how to construct and how to use with the battery in order to obtain the greatest effect. Faraday made his principal discoveries in magneto-induction in ten days when he was at the age of forty-two. There is little doubt in my mind that other men would speedily have discovered the same phenomena, for their universality could not long have eluded observation. Faraday's great achievement was in his conception of the lines of force which emanate from a magnetic pole and stretch through the ether of space; in his pointing out that the medium surrounding the wires carrying electric currents, and the medium in which magnets are situated, is in a state of strain; that there is what he called an electrotonic state of this medium. It was like a mass of quivering jelly—any movement at one point produced a quiver in all neighbouring points.

Maxwell, in his great work on electricity, thus speaks of Faraday's conception: "Faraday saw lines of force traversing all space where the mathematicians saw centres of force attracting at a distance. Faraday sought the seat of the phenomena in real actions going on in the medium; they were satisfied that they had found it in a power of action at a distance impressed on the electric fluids." * One obtains a realizing sense of the intellectual power of Faraday by reading his Experimental Researches in Electricity. He is not merely an inventor who, having discovered some phenomenon of Nature, proceeds to put it to a practical use; but each experiment is guided by a remarkable generalizing faculty, and the series of his experimental researches led him to the conception of electrical actions in the medium of space and laid the foundation of the greatest generalization in science of modern times—Maxwell's electromagnetic theory of light. We shall be led to this great theory as we continue our study of the question, "What is electricity?" Meanwhile let us examine a little further the phenomena discovered by Faraday.

A delicate galvanometer reveals, we have seen, that the motion of a wire near a magnet results in an electric current in the wire. Why this is so we do not

---

* Preface to Maxwell's Electricity and Magnetism.

know. We have a good working theory, however, which we shall express later.

Since the earth is a magnet, any motion of a wire on its surface will cause a current in the wire. One can signal under the sea through a cable by properly waving a coil of wire in the air. We say that the resulting current of electricity is due to cutting the lines of magnetic force of the earth by the motion of the wire.

For forty years, nearly half a century after Faraday's great discovery of magneto-induction, men's minds were almost exclusively devoted to obtaining steady currents of electricity in one direction instead of to-and-fro currents, such as are obtained by rapidly thrusting a north pole into a spool of wire and rapidly withdrawing it. The commutator was improved in every possible way until the commuting of the directions of the to-and-fro currents had well-nigh become perfect. The commutator in the best forms of the modern dynamo machine shows very little sparking, whereas in the earlier forms there was a brilliant coruscation of sparks when the segments of the commutator ran under the brushes which collected the current for the outer circuit in which the electrical work was to be done. These sparks showed that energy was lost. There is very little room for improvement at present in the modern dynamo. It approaches nearer to perfection than any other machine which is used to transmit power. It is doubtful in my mind whether Faraday ever realized the powerful effects that could be obtained when his lines of magnetic force were made to quiver with great speed. The galvanometer employed by him could only detect steady currents, or momentary currents which reversed in direction very slowly. It was incapable of showing any effect when the currents of induction were sent to and fro through it very rapidly. It would remain perfectly quiescent, its little needles pointing placidly north and south, while to-and-fro currents of tremendous energy were circulating through the circuit with which it was connected. It was like a deaf-mute with respect to the world of harmonies of an orchestra.

Although these to-and-fro currents of electricity annulled each other's effect on the needle of the galvanometer, they could produce an electric light, could heat wires, and, in short, produce all the effects obtained from steady currents with the exception that they could not run an electric motor of the type which we have considered—the type which had been slowly perfected during the forty years after Faraday's discoveries. We are now entering upon another period of electrical invention which may be called the period of adaptation of Faraday's discoveries to instruments adapted to to-and-fro currents instead of steady currents, and we shall see the necessity of using to-and-fro currents instead of steady currents for the transmission of electrical power, as we continue our study. I have said that it is questionable in my mind whether Faraday realized the wonderful development which is now beginning in the commercial employment of to-and-fro currents. He certainly, however, had a full conception of the sensitiveness of the electrotonic state of the medium surrounding magnets and electrical circuits, but he had no instruments which could represent to other people's eyes and ears the wonders of his imagination.

The telephone is an instrument based entirely upon Faraday's discovery of magneto-induction. If we should take the spools which we have used (Fig. 10) to illustrate the action of a commutator, slip each of them upon the poles of a magnet, as in Fig. 12, connect the ends of the wires of the spools permanently, place a thin disk of iron, such as is used in taking tintypes in photography, very near to each pole, providing a suitable earpiece to each telephone, we would find that a mere tap with the finger on the iron disk of telephone A, for instance, can be heard, on listening at that of telephone B, even when the telephone B is

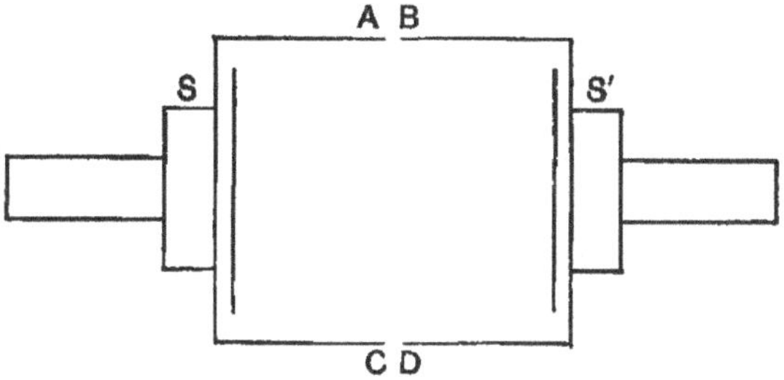

FIG. 12.

at the distance of many miles from telephone A. The operation of this tap is like thrusting rapidly a magnetic pole into the spool of B and quickly withdrawing it: to-and-fro currents of induction are produced which attract and repel the disk of the telephone B, thus reproducing the mechanical action which is exerted by tapping the instrument A. It is more correct as well as more picturesque to say that the quivering of the magnetic lines of force due to the slight movement of the thin iron disks affects the electrotonic state of the medium in which the telephones are immersed. The telephone into which we speak corresponds to the electric dynamo, and the telephone to which we listen to the electric motor. But here there is no commutator; we are using to-and-fro currents, or, in ordinary practical language, we have an alternating current dynamo and an alternating current motor. The excursions or movements to and fro of the telephone diaphragm are exceedingly small, and can only be detected under the microscope. Indeed, the telephone disk provided with a suitable pointer, in conjunction with a microscope, has been suggested as a suitable alternating current galvanometer, and Prof. Rubens, of Berlin, has lately shown how a practical instrument can be made on a similar principle.

If now we cause a sufficiently powerful alternating or to-and-fro current to circulate through a coil of wire wrapped around a piece of iron, or, in other words, if we use the big magnet employed by Faraday and Henry in their researches, we find that, although the galvanometer is absolutely quiescent and gives no indication of the great fluctuations of magnetism in the magnet, a telephone connected to a little coil of wire which is placed between the poles of the electro-magnet gives forth a loud hum, which is the note of the periodic currents; it is their rate of alternation. Furthermore, by moving the little coil about the poles of the magnet we can trace the spreading of the magnetic lines into the air, and thus through the ear obtain the same conception of this spreading as we obtain when we sprinkle iron filings on a piece of paper and place the paper on the poles of the magnet. Moreover, it is not necessary to connect the telephone with a little coil of wire; we can even remove the coil from the interior of the telephone, and thus with nothing but a thin disk of iron placed close to the pole of a magnet, by placing this dissected telephone, so to speak, to the ear, we can hear the hum of the electro-magnet in all parts of a large room, thirty or forty feet from the magnet, and by walking about the magnet we can trace the paths of the lines of force by the sense of hearing. We thus perceive that the entire space of the room is filled with these quivering lines of force. To-and-fro currents thus appeal to another sense—that of hearing—and greatly enlarge our conception of Faraday's electrotonic state. The lines of magnetic force crowd together into the thin disk and the magnet which form our exploring instrument. They prefer to pass through iron or steel to passing through the air. Thus, by placing a bar of iron across the poles of the electro-magnet we cease to hear the hum of the magnet in our exploring telephone. The lines of force prefer to pass from pole to pole through the bar of iron to spreading out into the room.

Now, if we could fill a room with these invisible lines of magnetic force of sufficient strength we could light an electric lamp in every part of it. The lamp would light when we entered the magnetic field, and go out when we retired from it. No matches would be required. We can realize this ideal system of lighting already in a small way, which I have often employed in my lectures to illustrate lighting up, so to speak, the magnetic curves, such as we see depicted in arrangements of magnetic filings. Connect a small coil of a suitable number of windings with a little incandescent lamp, such as is now used for lighting houses, but much smaller, move this coil near the poles of an electro-magnet through which a powerful alternating current is flowing—the lamp will light in certain positions of the coil in mid air. Moreover, we can move it about the poles of the electro-magnet so as to illustrate the curving of the lines of force. In order to fill a room with suitably strong lines of magnetic force to enable us to light a lamp in any part of the room by these invisible lines of force, we should need enormous magnets and enormously strong lines of force. Niagara Falls, however, is competent to-day to enable us to realize this flight of our imagination. It is interesting to observe that we could move about in this great field of magnetic energy with no sense of discomfort save from the hum of the alternating currents in the electro-magnet. There would be no sense of strain or pressure in the head. If it were not for the hum, we should not be conscious of the tremendous energy in the space of the room.

But I hear some one exclaim, "Are not certain people sensitive to magnets?" This belief has been held at various times by many persons, and after suitable interment it rises again to perplex humanity. On this subject one should consult Baron Reichenbach's treatise on what he terms the odic force (see page 40 of this treatise).

The future development of the practical employment of electricity in lighting and the transmission of power must be in the direction of alternating currents instead of steady currents. We shall see that Nature sends us, so to speak, our electricity by means of to-and-fro movements from the sun; and in order to approximate to the economy of Nature, we must adopt periodic movements, so to speak, of electricity in our apparatus.

Faraday's conception of the electrotonic state in the medium surrounding wires carrying currents and magnets is thus made definite by the quivering of the magnetic lines of force. These lines stretch out from the poles of a magnet and seek the shortest passage from one pole to another. They pass from the north pole of the earth to the south pole through the atmosphere, and every piece of iron or steel tends to turn so that the greatest number of these lines shall pass through its substance. The oxygen of the air itself is magnetic, and more lines of force will pass through it than pass through copper, for instance. Faraday believed that all bodies are more or less magnetic, and this has come to be the modern conception. The lines of magnetic force pass through all substances to a greater or less extent. We can conceive of a ship's compass turning so as to receive the greatest number of the magnetic lines of the earth, and therefore pointing to the poles of the earth. The magnetic lines of force which stream from one pole of the earth to the other are diverted in a thousand directions, passing into the gas pipes and water pipes of our cities, through iron-bearing strata of the earth, through the iron plates of the ocean steamships, so that it is difficult to find even a small space on the earth's surface where the lines of magnetic force are parallel to each other and of the same intensity. Instead of making the lines of magnetic force quiver, we can keep them steady and produce electric currents by cutting them, so to speak, by wires. When we revolve a coil of wire near the pole of a magnet we produce electric currents in the coil. These currents can be said to be produced by withdrawing lines of force from the coil or putting them into the coil. They can also be said to be produced by the turns of wire constituting the coil cutting the lines of magnetic force. If we should take in both hands a wire which is connected with a galvanometer and move it rapidly across the pole of a strong magnet, thus cutting the lines of force which emanate from the magnet, we would produce a current of electricity in this wire. Why, we do not know. We could have moved the magnetic pole instead of the wire, and could thus have produced the same effect. Electrical currents therefore result from the relative movements of magnets and wires. We can even light a room by merely revolving a large coil of wire with great velocity, so that it should cut the magnetic line of the earth at each revolution. The coil, however, would need to be so large and the speed so great that it would be better to revolve a small coil between the poles of a powerful electro-magnet, as is done in the case of the dynamo machine. The electro-motive force of the current obtained by cutting the magnetic lines of force is proportional to the number of lines of force that is cut each second. This electro-motive force can be thought of as the electric pressure which is obtained on the wire through which the electric current flows. The slightest movement of a wire anywhere on the earth's surface can be said to produce an electro-motive force in the wire, if such a movement cuts the magnetic lines of force of the earth. We therefore see the importance of arranging the pole pieces of our electro-magnets in order that we can obtain the greatest number of lines of force in the space between them, and so that our revolving coils which are to cut these lines of force shall cut the greatest number possible.

Instead of using strong permanent magnets to produce powerful magnetic fields, we coil insulated copper wire around bars of soft iron and pass powerful currents of electricity through the coils, thus forming an electro-magnet. It is easy to see how greatly we can increase the number of lines of magnetic force in this way, and it is instructive to trace the building up of a strong magnetic field by means of an electro-magnet.

If we bore a number of holes in a sensitive photographic plate and pass a wire through them so as to form a coil, and then place a piece of iron in this coil, we have thus an electro-magnet. On sprinkling iron filings on the sensitive plate, exposing the latter to the light, we see that the circular lines of force around the wire crowd through the iron core of the coil, making it a powerful magnet. The greatest number of lines of force we can urge through a square centimetre of soft iron is not far from thirty thousand. When a large number of lines of force pass through the air from one pole to another of a magnet an appreciable strain upon the air is noticed. It seems as if there was a pull or tension along the lines of force and a pressure at right angles to them. This pressure at right angles can readily be shown by suspending a glass rod between wedge-shaped pole pieces of iron. The magnetic lines crowd across in the air between the wedges and the glass rod moves aside to a place of less pressure.

Since in dynamos and electric motors it is necessary that the magnetic lines should pass from one pole to the other of the electro-magnets between the poles of which revolve the coils of the armature, and that few lines should be lost by diverging into the air, we see that watches should not catch stray lines of force, even when quite near the modern dynamo. If one's watch should be magnetized while one is riding in an electric car, it would be a proof that the electric motor propelling the car is a very uneconomical one. The shields of iron which are said to protect certain watches from being magnetized are generally inoperative, for the lines of magnetic force are not entirely diverted from the steel springs or detents of the works of the watch. In order to be effective, such shields should be of very soft iron, nearly half an inch thick. No substance cuts off the lines of magnetic force; they pass through wood, brick walls, copper, and all metals. They prefer to pass into soft iron on their way from one pole to another, and therefore iron answers as a species of shield. For instance, a galvanometer placed entirely within a thick shell of soft iron is used on shipboard, in laying submarine cables, for the purpose of testing the cable. The little magnets of the galvanometer are unaffected by the magnetic condition of the machinery of the vessel, and only respond to the electric impulses which are sent through the coils of the galvanometer. The reason is that the magnetic lines of force pass from the space outside the iron shell into the walls of the shell, and then out again without crossing the air space inside the shell, where the galvanometer is placed; and none of them pass through the needle of the galvanometer. Of course it would be worse than useless to try and protect the ship's compasses from the magnetism of the ship by inclosing them in iron shells. The lines of force of the earth would be diverted from the needle around the shell, and the needle would no longer point north and south; it would be useless as a compass.

If we could cut off the lines of magnetic force by some nonmagnetic substance, perpetual motion would be possible, for it would be merely necessary to place a magnet on a wheel in the direction of a diameter; bring a powerful magnet near one pole of the magnet on the wheel; set the wheel in rapid revolution, and at the instant the pole of the magnet on the wheel approaches the pole of the magnet which is placed outside the wheel insert automatically the supposititious shield. The magnetic attraction being cut off, the wheel carries the magnet around by its inertia again to the sphere of attraction, and the magnetic force is again cut off, and so on.

The steady lines of magnetic force can not be cut off

by copper, but if we make these lines quiver or alternate in direction we shall find that copper intercepts them. If, for instance, we place a sheet of copper between the pole of an electro-magnet through which a powerful alternating current is passing and a little coil connected with an electric lamp; the lamp is extinguished the instant the copper sheet is interposed between the coil and the magnet. In a little while we find that the sheet of copper becomes very hot, and we therefore see that energy has disappeared in the copper. We find by experiment that little electric currents are formed in the mass of the copper by the quivering of the lines of magnetic force which emanate from the electro-magnet, and these currents thrust magnetic lines into the field opposed to those coming from the electro-magnet, and thus neutralize their effect on the little coil of wire. We know, in the case of steady lines of magnetic force, that the movement of a copper wire across these lines of force will produce a current of electricity in the wire; and that if we keep the wire still and make or break the current of the electro-magnet, we shall also produce a current in the wire. We should expect, therefore, that a current would flow through the particles of copper of a plate. When we cause the lines of magnetic force to quiver or alternate, every particle of copper may be said to cut the lines of force. The entire phenomenon is one of the excitation of an electric current by the movement of a conductor across the lines of magnetic force or the movement of the lines of force across the conductor. In each case a current of electricity is excited in the conductor. If we should suspend by a fine fibre a light magnetic system, made by magnetizing two cambric needles, sticking them into a match so that their opposite poles should lie over each other, and twist the strand by which the system is suspended, it will come to rest sooner when it revolves over a copper plate than it does when it revolves far from all neighbouring objects. The poles of the magnet nearest the copper plate induce currents of electricity in this plate, and the lines of magnetic force resulting from these currents in the plate tend to pass into the revolving magnet. The latter desires to receive them, and then tends to come to rest.

In all these cases there is a decay of energy in the copper. If we should look upon the alternating electro-magnet as a source of undulations or electro-magnetic waves which it sends out into the surrounding space, we see that such undulations can not pass through what we call good conductors, such as copper. If we take away the copper plate, which shields a coil from an electro-magnet through which an alternating current flows, and substitute a glass plate a little lamp connected to the coil will light, for there is no decay of energy in the glass; it allows the electro-magnetic impulses or undulations to pass freely through it. With reference, therefore, to very rapid electro-magnetic undulations, we shall see that our present nomenclature is wrong. To such undulations copper is an insulator and glass a good conductor. Copper absorbs the energy of the undulations, and glass does not.

Our attention, therefore, has been directed to the distribution of lines of force in and about magnets and wires carrying currents of electricity. We have in imagination filled space with such lines which represent the flow or stream, so to speak, of what is called induction. At first sight it would seem that we have returned to fluid analogies, and that we are about to proclaim that all electro-magnetic phenomena are due to the flowing of the ether into and out of magnets or around wires carrying currents, and that we are ready to assert that electricity is a motion of the ether. We shall see, however, that such an assertion is certainly premature, and that at present we can merely regard the conception of lines of force and flow of magnetic induction, which, in other words, is the number of lines of induction across any surface or through any coil of wire, as an aid to our calculation of the mechanical effects observed.

The conception of the flow of magnetic induction is of great use in our classification of the vastly complicated efforts which we observe in the subject of electricity. We say that the north and south poles of the magnets attract each other, because each pole tends to embrace the greatest number of lines of force. Two circles of wire through which currents of electricity are flowing in the same direction turn their planes parallel to each other, in the same endeavour to embrace the greatest number of lines of induction. They are attracted toward each other in this effort. In fact, these circles act exactly like magnets. If the currents flow in opposite directions in the two circles they repel each other, or, in other words, they endeavour to turn their other face around so as to embrace the greatest number of lines of induction which are flowing in what we call a positive direction. If we place soft iron in the centre of our little coils of wire which are traversed by electric currents, we magnetize the iron—we create a flow of magnetic induction through the iron. In a permanent magnet, such as is used in ships' compasses, this flow of magnetic induction is always present, and, to fix our ideas, we may perhaps conceive of it as a streaming of the ether through the iron. If we place two north poles of two magnets near each other they repel each other, they tend to demagnetize each other, and the lines of flow between them is greatly weakened. If the magnets are free to move, it is well known that they will turn so as to present a south pole to a north pole; or, as we say now, they will turn so as to embrace between them the greatest flow of induction. If we endeavour to turn the poles away from each other we must do work. A force resists our effort to separate a north pole from a south pole. We find that this force is proportional to the product of the strength of the two poles divided by the square of the distance between them. In the same way we must do work to separate two parallel coils of wire through which currents of electricity are flowing in the same direction. These parallel coils present their north and south pole to each other also. They are electro-magnets, and the force we must overcome in separating them is proportional to the product of the strength of the currents which are circulating in them. In general we find that magnets and coils of wire in which electric currents are circulating tend to turn so as to embrace the greatest flow of induction between them.

The principle we have endeavoured to elucidate in the last paragraph is one of the most important in the subject of electro-magnetism, taken in connection with the fact that any change in the flow of induction through coils produces currents in these coils. For instance, we know that the fluctuation of a current in one coil will produce a momentary current in a neighbouring coil; and the force between the coils is proportional to the product of the two currents, and tends to prevent the lessening of the flow of induction between the coils.

A simple method of showing that the induction current in an induction coil on making the circuit of the primary is opposite in direction to that in the primary, and that on breaking the circuit it is in the same direction, is as follows: Make a file cut at the centre of a light brass rod and balance the rod on the edge of the blade of a pocketknife, having previously hung two scale pans made of bristol board on the ends of the rod (Fig. 13); twist the ends of a piece of copper wire together and bend the wire into the form of a ring; place this ring in one scale pan, having counterbalanced it, and bring it over the end of an electro-magnet — the latter serves as the primary coil, and the ring as the secondary coil; on making the circuit of the primary the ring will be repelled from the electro-magnet, and on breaking the circuit it will be attracted. In the first case two similar poles are formed, and in the second case two opposite poles. By placing a little mirror on the balance arm at K, and by the observation of its movements by means of a telescope and scale, a very feeble battery may be employed to excite the electro-magnet. Theoretically, a copper ring suspended above the pole of the permanent magnet of a receiving telephone should vibrate when one speaks into a sending telephone. Indeed, telephone diaphragms have been made of parchment with small closed circuit

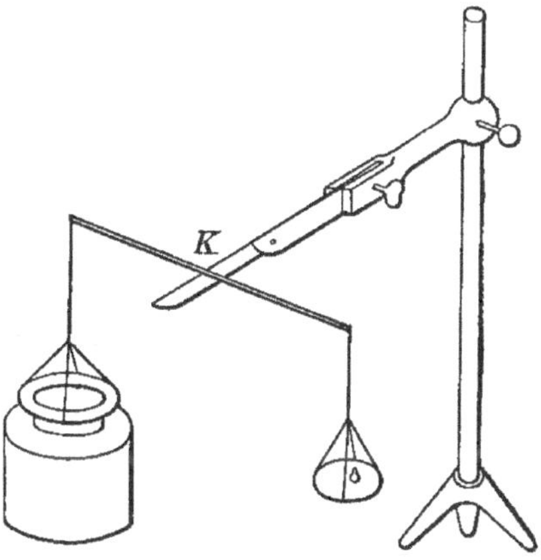

FIG. 13.

coils glued upon them opposite the pole of the receiving telephone.

In 1880 the continuous-current dynamo may be said to have reached its point of perfection. Its efficiency is as high as eighty-seven per cent, and can be made under good conditions as high as ninety per cent—that is, there was only a loss of ten per cent in the transformation from steam power to the electrical current. The dynamo is well-nigh perfect. The distribution, however, of the current produced by it is far from economical if such distribution is extended to considerable distances. It is found that a light can not be produced at a distance of ten miles from the dynamo without great loss of energy along the wires leading to the lamp. This energy could be saved by using massive copper conductors instead of the ordinary copper wires which are employed at short distances, but the cost of such great conductors is prohibitive. In ordinary language, it is said that the electric pressure diminishes very rapidly with the distance; and in the ordinary electric railway, such as that of the West End Railway in Boston, Mass., it is found necessary to feed the overhead wire at various points by auxiliary conductors in order to keep up the electric pressure, or potential. In order, therefore, to substitute electric power for steam on a railway between Boston and New York, it would be necessary with a continuous current to have power stations every ten miles. For considerable distances electrical power is hardly more economical than compressed air, and very high authorities pronounce in favour of compressed air.

Since the dynamo has arrived at a high degree of perfection the attention of inventors is now turned to the important question of the more economical distribution of the current generated by the dynamo, and in the consideration of the new methods of distribution we are led to a remarkable development of electrical science—that of the varied uses of the alternating current. With this latter current, which flows to and fro, changing in direction sixty to a hundred times a second, like the currents in a telephone—although the alternations produced by the voice are much faster than this rate—a hundred horse power has been transmitted one hundred and ten miles from Lauffen, on the river Neckar, to Frankfort by means of three wires, each about one sixth of an inch thick. The loss of power in this transmission was barely twenty-five per cent, far less than would be the loss in transmitting the same amount of power five miles by a continuous current. It is proposed to use the alternating system in the transmission of power from Niagara Falls. It would seem as if our mechanical contrivances to direct the alternating currents which are produced by every dynamo in the same direction by means of a commutator were working against Nature, and that we are returning to a manifestation which is analogous to that of the electro-magnetic waves in the ether. If we should dispense with a commutator, and place instead of it two rings on a revolving shaft to which the wires of the coils which are revolving past the poles of a magnet are connected, we should collect from these rings by suitable brushes an alternating current. The modern alternating machine, therefore, consists of stationary electro-magnets fed from a continuous-current dynamo, with its commutator, and a revolving armature the coils of which are connected to two rings on the revolving shaft. We need the commutator on the dynamo which is furnishing the current for magnetizing the field magnets of the alternating dynamo. We do not need, however, a commutator to take off the alternating currents produced by the revolution of the armature coils of the alternating dynamo. Now, the principal reason why the alternating current is superior to the direct current in the transmission of power to a distance resides in this: that with the alternating current we can transform a comparatively low electric pressure or voltage to an enormously high one, and so overcome the electrical resistance of a long line of comparatively fine wire.

This transformation is accomplished by means of an apparatus which is substantially the Ruhmkorff coil. This consists of two parts an electro-magnet wound with a few turns of coarse wire with a core of bundles of fine wires, and a fine-wire coil entirely separate from the coil of the electro-magnet, but wound closely upon it. When a battery current is made and broken through the electro-magnet coil, which is termed the primary, a flow or flux of magnetic induction is sent through the fine-wire coil; a strong electro-motive force or pressure is set up in this coil, since we know that the flux through a coil always produces a difference of electric potential in the coil. With a potential of only four units or volts in the primary, we can obtain at least ten thousand volts in the secondary circuit.

CHAPTER XI.

ALTERNATING CURRENTS.

We can prove by the doctrine of the conservation of energy that the direction of the induced current in a coil neighbouring to that carrying the inducing current will be opposed in direction to this latter current. For, suppose that the current in the coil on the circuit A B (Fig. 14) induces a current in the coil on the circuit C D in the same direction as itself, as shown by the arrow; on connecting the circuit A B with the circuit C D, as indicated by the dotted line, we should be able to increase the current flowing through both coils by placing them parallel to each other, for we should have the induced current superimposed on the inducing current, and in the same direction. We should get an increase of energy merely by placing coils parallel to each other without doing any work. That is, we could get more out of a battery which is supplying the current by placing the coils in the circuit near each other, than we could by placing them apart.

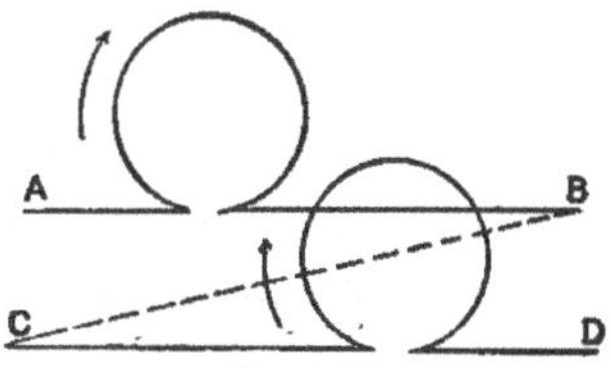

Fig. 14.

The doctrine of the conservation of energy can also be applied to the determination of the direction of the currents which are produced by moving a coil in a magnetic field. Let us take the case of the movement of a coil across the face of the pole of a magnet. When we draw the coil away from the pole a current is induced in the coil. This current is due to the change of the flow of induction through the coil; or, as we ordinarily say, it is due to the removal of lines of force from the coil. Now the induced current must flow in such a direction that the attraction of the pole formed by it—the coil becoming for a moment an electro-magnet—on the pole of the magnet must tend to resist the movement of the coil; for if the pole of the electro-magnet repelled the pole of the magnet, we should be assisted in moving the coil away from the magnet, and this would be contrary to the conservation of energy, for we should be gaining more work than we are doing. Let us now see what would happen if, instead of moving a coil from a magnet, we should move the magnet away from the coil. In this case, also, we see that the current induced in the coil must be in such a direction as to resist the movement of the magnet, and if the coil were properly suspended it would follow the magnet. A simple way of trying the experiment is as follows: Suspend a copper disk by a single thread over a magnet which can be made to revolve close to it and under the disk. As the poles of the magnet sweep under the disk currents of electricity are induced in the disk, just as if the disk was made up of coils of wire. These currents flow in such a direction as to oppose the movement of the magnet. They form little electro-magnets, the poles of which attract the poles of the moving magnet, and consequently the disk is swept round with the magnet. In this simple apparatus we have made our first acquaintance with the rotary magnetic field, which is becoming of such importance in the problem of transmitting power over great distances. The copper disk is the armature of our motor, but we shall see later that it is not necessary to use a rotating magnet. We can produce the same effect by alternating an electric current suitably through electro-magnets which are fixed beneath the disk.

In order to understand the action of fluctuating or alternating currents upon each other, we must consider what is termed the phase of the currents with respect to each other. It is difficult to obtain a clear idea of this word phase, but it is necessary that we should do so.

If, in Fig. 15, *a b c d e* represents a wave, *c* being the crest of the wave and *d* being the trough, and also

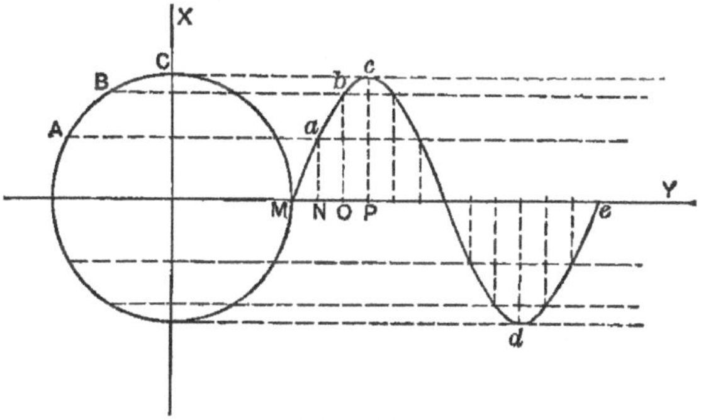

FIG. 15.

if *c* and *d* represent two boats, *c* will be at the height of its upward movement when *d* is at the lowest point. The difference of phase of the boats in regard to their relative moment is said to be 180°, for if we draw a circle, A B C M (Fig. 15), we can represent the motion of the boats by the relative movement of two points around the circle. The rise of A to C and the fall of M to the point on the circle opposite C in the same time can be represented by the curve *a b c d e*, in which the distances along M Y represent the times when A and M have reached different points in their run around the circle. As another illustration, suppose we have two pendulums, one of which, C, Fig. 16, is passing through the middle point of its swing, while the other, D, is returning, having reached the extreme amplitude of its swing. These pendulums have a difference of phase of 90°, and their motions can be represented also by the movements of two bodies around a circle. If the boats *c* and *d* were together, they would rise and fall together and would be in the same phase.

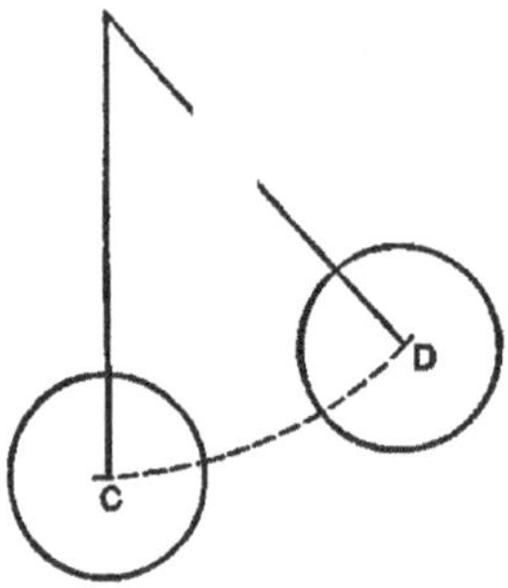

FIG. 16.

As a practical illustration of difference of phase, let us examine the working of the telephone. Is there any difference of phase between the motion of the diaphragm against which we speak and that of the diaphragm to which we listen? The human voice sets the iron diaphragm in motion, and when it is moving swiftest it is causing the greatest disturbance of the lines of induction between it and the magnet of the telephone. The diaphragm of the receiving telephone starts into movement at the instant of this swiftest movement of the sending diaphragm. It is therefore in the condition of a boat at M (Fig. 15) while the sending diaphragm is in the condition of the boat at *c*. There can be a difference of phase of 90° between them.

Let us now examine what takes place when we put a copper ring in front of an electro-magnet through which an alternating current is flowing. This alternating or periodic current produces a fluctuation in the flow of induction through the neighbouring copper ring, and an alternating current flows around the ring. If there were no lagging of the current in the ring, due to self-induction or the setting up of the lines of force in the surrounding medium, the conditions of movement would be represented in Fig. 17, in which the strong line represents the primary or inducing current in the electro-magnet, while the thin line represents the secondary or induced current in the copper ring. The difference of phase between C and D is 90°. Now if the height of the two lines above or below the horizontal line C M represents the strength of the currents at any time, we know that the force between the electro-magnet and the coil is proportional to the product of these currents. If, therefore, we should multiply together the respective distances of two points, such as C D or E F, above and

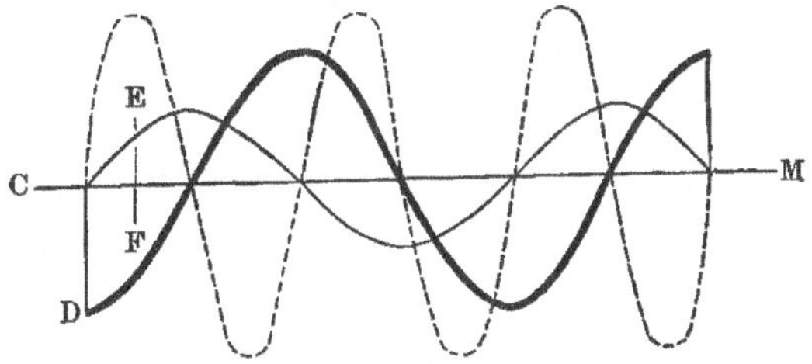

FIG. 17.

below the middle line C M, we should obtain the force of attraction or repulsion. The strength of the current of induction at C is nothing, and that of the inducing current is C D. Hence, nothing (or zero) multiplied by C D is nothing, and the dotted curve which represents the resultant attraction or repulsion starts from C, and rises and falls in the manner indicated by the dotted line. We should have alternate equal attractions and repulsions. "Currents would be induced in opposite directions to that of the primary current when the latter current was changing from a zero to maximum positive or negative current, so producing repulsion; and would be induced in the same direction when changing from maximum positive or negative to zero, so producing attractions." * The ring, therefore, would not be attracted or repelled by the electro-magnet. The current, however, in the copper ring is not in phase with the current in the electro-magnet or the inducing current.

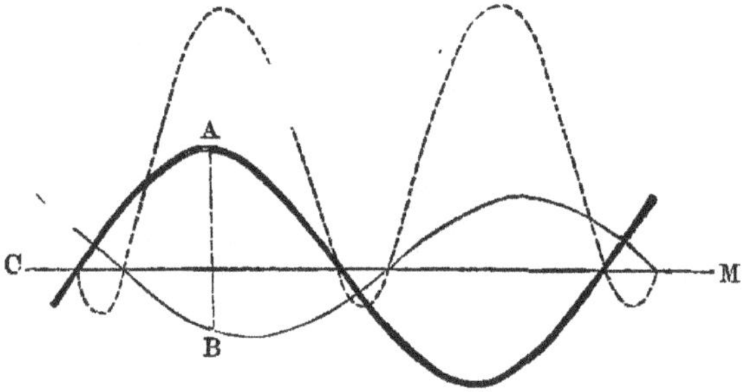

FIG. 18.

The reason that it is not in phase or in step with the inducing current is due to the work that it must do in establishing lines of force about itself—in directing, so to speak, a flow of induction around its circuit. The time that is spent in this work is different from that spent in the inducing circuit in similar work. The crests and troughs of the waves in each circuit, therefore, can not be represented by the thick and thin lines of Fig. 17, but are represented by the lines of Fig. 18. The dotted line is obtained, as before, by multiplying

* Prof. Elihu Thomson, Novel Phenomena of Alternating Currents.

together the strengths of the current represented by the distances of any two points, such as A and B—one on the thick curve and one on the thin curve—from the central line C M. "During the period of repulsion both the induced and inducing currents have their greatest values, while during the period of attraction the currents are of small amounts comparatively. There is then a repulsion due to the summative effects of strong opposite currents for a lengthened period against an attraction due to the summative effects of weak currents of the same direction during a shortened period, the resultant effect being a greatly preponderating repulsion." A copper ring suspended opposite the pole of an electro-magnet through which circulate alternating currents is repelled. I have been accustomed in my lectures to show the following modification of this experiment: A thin copper cylinder is slipped over a number of pieces of soft iron wire which protrude from the centre of an electro-magnet. When an alternating current is suddenly sent through the electro-magnet, the iron wires fly into the centre of the coil, while the copper cylinder is shot in the opposite direction many feet.

We owe to Prof. Elihu Thomson a number of similar interesting experiments. If, for instance, a copper ring is held over an electro-magnet through which is circulating an alternating current, it is repelled, and can be made to float in the air without visible means of support. A thin copper sphere suspended over the pole of the electro-magnet, one half of which is shielded by a plate of copper, immediately begins to spin. Its rotary action is due to the difference of phase between the currents induced in the copper plate and in the sphere. Prof. Thomson observed that if a copper ring was held in an oblique position in front of the pole of an electro-magnet, it tended to turn so as not to inclose any flux of magnetic induction. He therefore took a continuous-current armature, such as is employed in the continuous-current dynamo, and connected the brushes by a wire. When an alternating current passed through the field or stationary magnets of the dynamo, the coils of the armature acted like the copper ring and tended to place themselves so as not to embrace the flow of magnetic induction. This arrangement constituted an alternating-current motor.

It is interesting to notice that Faraday observed this phenomenon of the turning of the copper ring, and Maxwell, in his Electricity and Magnetism, remarks: "Hence the effect of magnetic force on a perfectly conducting channel tends to turn it with its axis at right angles to the lines of magnetic force—that is, so that the plane of the channel becomes parallel to the lines of force. An effect of a similar kind may be observed by placing a penny or a copper ring between the poles of an electro-magnet. At the instant that the magnet is excited the ring turns its plane toward the axial direction, but this force vanishes as soon as the currents are deadened by the resistance of the copper." *

These interesting experiments of Elihu Thomson I found, to my surprise, could be performed with very simple apparatus. Although they were discovered by the use of powerful alternating currents, they can be repeated with two or three cells of a battery, the current of which is made and broken through an electro-magnet by means of an ordinary interrupter, such as is commonly used on a Ruhmkorff coil and known as the hammer-and-anvil

* Maxwell's Electricity and Magnetism, vol. ii, § 843.

interrupter. The wonder is that these remarkable phenomena should have escaped observation so long. I arrange the apparatus as follows: The coarse coil through which the battery current is rapidly made and broken by the interrupter is first placed in a horizontal position, having been provided with a bundle of fine iron wires for a core. Another smaller bundle of fine iron wires (B, Fig. 19) is held by a suitable clamp directly opposite the end of the electro-magnet. On

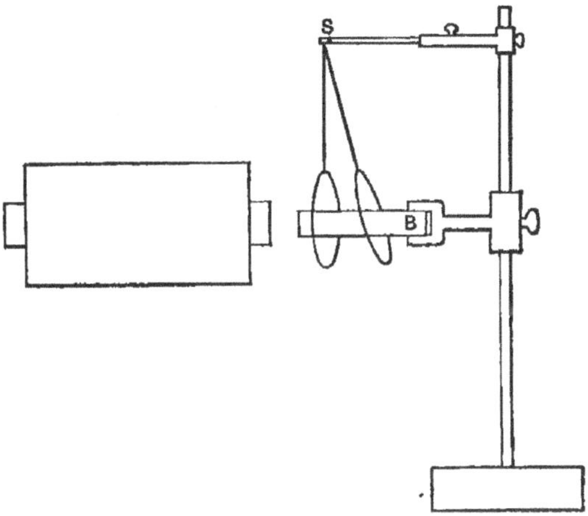

FIG. 19.

this latter bundle is slipped a bare copper wire, of which the ends are twisted so as to form a light ring. This ring is suspended by a loop of silk thread from a support, S. When the current is started in the electro-magnet the little ring is instantly repelled along the iron core B, and endeavours to set its plane parallel with the bar B. A heavier copper ring is more suitable than the very light ring to show the turning action. The electro-magnet is then placed in a vertical position (Fig. 20), a thin copper disk is suspended over one half of the pole of the magnet, and a copper plate is placed partly beneath the disk. When the magnet is excited the disk is set in rapid rotation.

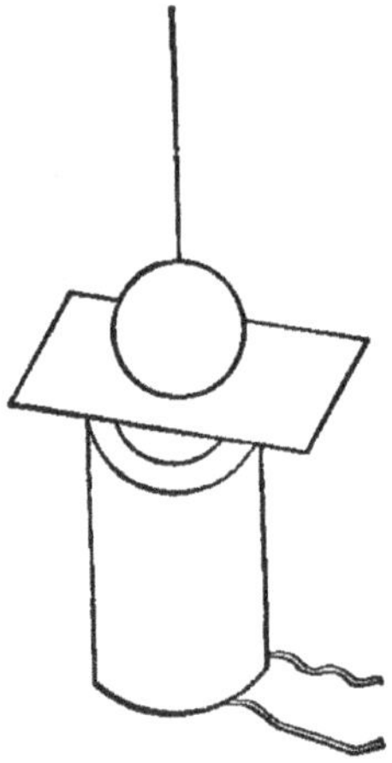

Fig. 20.

An interesting modification of the latter experiment is, to substitute a hollow light brass ball, such as are used as ornaments on certain curtain fixtures. The ball rapidly spins about its axis of suspension.

The copper ring in this experiment gets very hot if strong alternating currents are used. If it is suitably placed in a vessel of water, the water can be made to boil. We have here a transformation of electric energy into motion and also into heat. The heating effect can also be shown by the employment of very simple means within the reach of almost any experimenter. To the

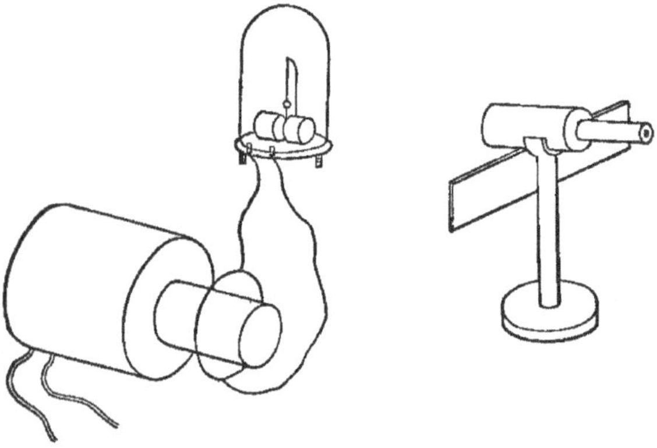

Fig. 21.

copper wire which we have used to show the effects of repulsion, solder an iron wire at one end of a diameter and a copper wire at the other end of this diameter; connect these wires to a suitable galvanometer (Fig. 21), slip the copper ring on the pole of the magnet, insulating it from it by a roll of paper; when the circuit is made and broken through the electro-magnet the galvanometer speedily shows that the copper ring is heating. We have in this case a thermal juncture of iron and copper on the ring and the other juncture outside the ring. Two cells of a battery will readily show this phenomenon. The ring is a step-down transformer of very small resistance, combined with a very small electro-motive force, and with consequently a comparatively large current.

CHAPTER XII.

TRANSMISSION OF POWER BY ELECTRICITY.

We have pointed out that the force of gravitation affords us our practical measures of electricity. It can also produce electricity by means of the weight of water.

In 1877 Sir William Siemens, in his presidential address before the Iron and Steel Institute of Great Britain, spoke of the possibility of utilizing the power wasted in the falls of Niagara, and said: "Time will probably reveal to us effectual means of carrying power to great distances, but I can not refrain from alluding to one which is, in my opinion, worthy of consideration—namely, the electrical conductor. A copper rod three inches in diameter would be capable of transmitting 1,000 horse power to a distance, say, of 30 miles." Again, in 1878, he states that there would be sixty per cent lost in transmitting this amount of power by electrical means over a distance of 30 miles.

In the year 1882 M. Depretz attempted to transmit power from Weissbach to Munich over 35 miles of iron telegraph wire 0·18 inch in diameter. He used a dynamo such as is commonly employed to-day on arc-light circuits, and the pressure which forced the electricity, in common parlance, along the wires was 1,500 units, or volts, as they are termed. It is important to notice the amount of this pressure, for in modifying this factor greater success has been reached. The increase in the pressure seems to be the key to the entire situation. The first experiment of Depretz was not entirely satisfactory, and it was repeated in 1883; but the second experiment was far from being successful. In 1883 still another experiment was made, which was far in advance of previous experiments. Power was transmitted from Vizille to Grenoble, in France, a distance of $8\frac{3}{4}$ miles, with a silicium bronze wire of 0·079 inch in diameter. Seven horse power was obtained at the receiving end, the loss being only sixty-two per cent. The improvement resulted from employing 3,000 volts instead of 1,500. Here it was clearly indicated that the direction in which to work was in employing high electrical pressure or voltage.

It was soon discovered that advance was barred in this direction of increasing the pressure by the impossibility of making a dynamo which would furnish the high pressure with a continuous current, such as is commonly employed to-day on our street arc lights. In a subsequent experiment M. Depretz endeavoured to construct a dynamo which would furnish a higher voltage; and although the experimental dynamo could not furnish the high voltage of 6,000 units which was desired and was burned out in the experiments, nevertheless M. Depretz showed that 52 horse power could be transmitted 35 miles over a copper wire 0·2 inch thick. Although this latter experiment was a failure, it clearly showed how greater success could be obtained. A higher pressure or voltage must be used, and instead of a direct-current dynamo a new type must be employed—namely, an alternating-current dynamo, one in which the electrical current pulsates to and fro, now in one direction and now in the opposite.

It will be noticed that during the years 1877–'83 men's minds had changed greatly in considering the subject of the transmission of power from Niagara Falls to even the distance of 30 miles. The early objectors to the scheme calculated the expense of the enormous copper conductor three inches in diameter, and showed the practical impossibility of the plan. The later objectors pointed out that, although power from the falls could be transmitted 30 miles over a wire only 0·2 of an inch in diameter instead of three inches in diameter, no dynamo could be constructed which could give the large pressure of 6,000 units and maintain its life. It would be burned out by the excess of its emotion.

The remarkable new developments, therefore, in the subject of the transmission of power by electricity come from the employment of a high electrical pressure or voltage generated by a to-and-fro or alternating current instead of a direct current; and, strange to say, an apparatus which has long been used on professors' tables to illustrate the conversion of a low-pressure current of electricity into a high-pressure current has now come to have a great commercial value; this apparatus is the Ruhmkorff coil. In its elements we have seen it consists merely of two coils of wire entirely separate from each other, which are slipped on a bundle of iron wire which forms a core. If an alternating or rapidly interrupted current of electricity from a battery or a dynamo is sent through one of the coils, a current is generated by induction in the neighbouring coil. The pressure in this latter coil depends largely upon the number of windings in it. A pressure of only two units in the coil which is connected with the battery or dynamo can be exalted to a pressure of 12,000 to 100,000 units in the independent coil by properly increasing its windings. Again, if the sparks from a Leyden jar which is charged to 100,000 volts by an electrical machine should be sent through a coil of many windings, a neighbouring independent coil can be made to give a pressure of only four volts, with, however, a large current for an instant. The Leyden-jar current is a very feeble one although it has a high pressure.

If it were possible to direct a bolt of lightning through a coil of many thousand windings of fine wire, a neighbouring coil of few turns of coarse wire entirely unconnected with the coil through which the bolt of lightning passes could be made to furnish a current for an instant which would decompose water into oxygen and hydrogen, although this would have been out of the power of the original bolt of lightning. In this system of exalting the pressure by employing induction in independent circuits—for instance, between two coils slipped upon an iron rod or bundle of wire—we have the modern transformer system, which marks a great development in the subject of the practical application of electricity. It will be noticed that we have a step-up transformer when we exalt a pressure of four volts to 100,000 volts, and a step-down transformer when we use a pressure of 100,000 volts in a fine-wire coil to produce a pressure of four volts in a neighbouring coil of coarse wire. A dynamo, therefore, producing only a 1,000 volt pressure and the dynamo ordinarily used to-day safely can stand this pressure, can be employed to send an alternating or pulsating current through the coarse coil which we slip upon our iron rod, and the fine-wire coil of greater number of turns which is placed near but unconnected with the coarse coil can be made to give by induction a pressure of from 12,000 to 20,000 volts. This pressure can be transmitted over a fine-wire one hundred miles, passed through another fine wire coil upon an iron rod, and a neighbouring coarse coil can be then made to yield a lower pressure suitable for decomposing water, running an electric motor without burning it up, or doing any work which the generating dynamo at the sending end is capable of.

In the ordinary use of step-down transformers for electric lighting it is customary to so arrange the ratio between the number of turns of wire in the primary to those in the secondary that an electro-motive force of a thousand volts in the primary is transformed to fifty or seventy-five volts in the secondary. A thousand volts is therefore present in the street circuit, and a current of only fifty volts enters one house. A thousand-volt circuit would be highly dangerous in a house, while one of fifty volts is comparatively safe. Such a system of transformation is far more flexible than the system of electric lighting by a direct continuous current, in which, of course, no transformers can be used, for the essence of the action of a transformer lies in the fluctuating flow of induction produced by an alternating or to-and-fro current of electricity. With a transformer one can obtain one light or a hundred or more, while with a continuous current one can not obtain one light advantageously without having a number lighted in the same circuit. By properly proportioning the transformers on a one-thousand-volt circuit one can range from the production of electric sparks to the production of an electric light of sixteen-candle power or to the production of the minute lamp that surgeons use to light up the human throat.

This great range illustrates the wonderful field of the transformation of energy which the study of electricity opens to us. It seems as if the nearer we get to the rate of pulsation of the electro-magnetic waves in the ether the nearer we get to an economical employment of this great source of energy.

Since we can obtain an abundance of light by a transformer, the question naturally arises, Can we not obtain great manifestations of heat by properly arranging the proportions between the primary and secondary coils? This has been done by Prof. Elihu Thomson in a remarkable process called electric welding. Suppose that the secondary wire consists of merely a copper ring of very small resistance, and that the electro-motive force excited in it is one volt or one unit: the approximate expression for the strength of the current in the ring is the electro-motive force divided by a very small quantity made up of the resistance of the copper ring and another factor depending on the rate of alternation of this current and what is called the coeffient of induction. One will therefore be divided by a small fraction. We may, for instance, have one volt divided by $\frac{1}{1000}$ and thus obtain a current of one thousand ampères or units. This current would speedily melt a bar of iron of the size of the average human wrist. Prof. Thomson by ingenious mechanism has made it possible to weld different metals together, so that the strength at the joint is superior to that at other portions of the rods. Metals can be welded together which can not be joined by brazing or soldering. The process enables one to apply great heat at exactly the point where we wish it. The two bars to be welded are brought together end to end. The principal resistance being thus at the imperfect contact of these ends, great heat is developed, the bars are raised to a white heat at their junction, and they are then pushed together.

One of the ingenious applications of this transformation of energy is to the welding together of shells used in warfare and to the annealing of steel armour plate in places where bolts or nuts are necessary. The steel plates are so hard that they resist the action of ordinary tools, and they therefore have to be softened in order to allow the boring for the bolts. The circuit of the secondary of the transformer can be closed at these points by resting the ends of the transformer ring on the iron. Great heat is therefore developed at such points, and the steel can be annealed at these points. It is evident that houses could be heated by a similar transformation, for coils of wire might be so arranged that currents of air in passing over them could readily be heated. This transformation at present is somewhat expensive.

In 1883 the attention of mankind was directed to producing dynamos which would give the steadiest continuous current; to-day we are striving to produce dynamos which will give to-and-fro currents, for the success of the new method of transforming electricity from one pressure to another depends upon the fluctuation of a current and not upon its steadiness. It may be said that our inventors unconsciously have been imitating Nature; for every bolt of lightning is not one continuous discharge, but is an alternating current which pulsates to and fro ten or twelve times, or even more, in a millionth of a second.

It is well known that the ordinary electric car is propelled by a dynamo which is similar to the dynamo at the central station, which generates the continuous current of electricity utilized by the dynamo motor in the car. The dynamo motor can be made to generate a continuous current of electricity if necessary. One dynamo can thus be said to be the counterpart of the other. The motor can be made the generator or the generator the motor. This is true, speaking in general terms. If we use an alternating-current generator, we can not use the ordinary motor such as is now used on electric cars. There must be a similar correspondence to that which characterizes generators and motors when continuous currents are employed; the alternating generator requires an alternating motor.

The recent success in transmitting power over one hundred miles is due not only to the method of transforming a current of low pressure to a high pressure and then transforming back at the receiving end, but also to the invention of an alternating motor. At first sight it seems impossible that an alternating current flowing through a dynamo could make the armature of the latter revolve, for the pole of the electro-magnets which attract the armature become alternating north and south poles and apparently neutralize each other's effect on the armature. Here we are brought again to see an apparatus which has long been exhibited on lecture tables developed into an important commercial engine.

Arago showed that a copper disk suspended by a thread above a revolving magnet would be dragged around by this magnet. Its motion was due to the reactions on the magnet of the currents formed in the copper by the motion of the magnet. Prof. Ferraris, in Italy, and Mr. Tesla, in America, showed that, instead of rotating a magnet, magnetic poles could be formed at different points in a collection of electro-magnets by means of alternating currents in such a manner as completely to imitate a revolving magnet.

The copper disk would therefore follow the changing poles. Instead of a disk an armature consisting of copper rods was made and the new alternating-current motor sprung into existence. Its most striking peculiarity consists in its absence of brushes, such as are used on ordinary dynamos. Its armature, or revolving part, is practically a copper disk revolving under exactly the same conditions as in Arago's experiment. To this or to a similar alternating-current motor we must look for aid in transmitting power great distances over a wire.

It is interesting to note that the conditions for the transmission of power by electrical means over long distances are closely analogous to those which are employed in long-distance telephony. The transmitter is an alternating-current machine operated by the human voice, and the telephone at the receiving end is an alternating-current motor which impresses its motion upon the air and thus reproduces the speaker's voice a thousand miles away. In long-distance telephony, too, the step-up transformer is used.

By the method of step-up and step-down transformers, which we have described, one hundred horse power has been transmitted one hundred miles from Lauffen to Frankfort over a wire which resembles in size an ordinary telegraph wire. Although the expense of providing copper conductors of great size (for instance, three inches in diameter) has been obviated, another difficulty now remains—that of properly insulating the line which is conveying a pressure of 12,000 to 20,000 volts. The higher the pressure the more tendency there is for electricity to escape from the line. In fact, a spark could be obtained by presenting one's knuckles to the wire charged to a pressure of 20,000 volts, and therefore the tendency of the current to leave the wire at each support and escape to the ground is very great. The wire between Lauffen and Frankfort was carefully insulated at the poles, carrying it by various devices, in one of which insulating oil was used. The pressure of 12,000 volts which was actually used between Lauffen and Frankfort was dangerous to human life. A skull and crossbones was painted on the poles which carried the wires, and no one cared to touch the wires. It is evident that if 12,000 to 20,000 volts are to be used on overhead wires in the future they must be as much respected as the way along which an express train travels.

It seems to me undoubtedly true that a diminution in our coal supply would result in making the transmission of power from Niagara Falls to New York a success. Mankind has often cast uneasy glances at these unemployed giants—waterfalls and tides—as if envying them their freedom. If we could make them store up their energy in a practical form, we should no longer be compelled to delve in mines a mile under the earth for coal, and we could fold our hands while the water ran on and did our work.

When the storage battery was invented it was thought that Niagara Falls had lost its freedom and would be immediately set to work. There was a period when the elation of mankind grew less and faith in storage batteries declined; for they were far from perfect and could not be relied upon. It was much as if the swift trotting horse had appeared before long experience had been obtained in training him and in properly nourishing him. The early experimenters endeavoured to drive the storage batteries too hard, and the battery broke down under severe usage. To-day the storage batteries are becoming a commercial success, and people understand better how to keep them in condition. There is therefore a possibility of employing them to convey a portion of the power of Niagara to Chicago. Let us see what this possibility is.

Roughly speaking, six horse power can be stored in a ton of material which constitutes the storage battery. The equivalent of fifty horse power could be certainly carried in one freight car, and it would therefore require one hundred freight cars to transport 5,000 horse power from Niagara Falls to Chicago. The cost of the batteries would be in the neighbourhood of $2,000,000, and in order to maintain 5,000 horse power in Chicago relays of batteries would have to be employed. Against the expense of this method we must place the cost of the high insulation of a line of four hundred miles under a pressure of 20,000 volts. In both cases, neglecting the factor of sublimity, it would be more economical to generate the electricity at Chicago from coal by the ordinary method of employing a steam engine to drive a dynamo. Improvements in constructing high insulation wires and in transforming high electrical pressures to low ones and, *vice versa*, may, however, make a greater revolution in the subject than has been made since the first experiments of Depretz, in 1882.

It is interesting to note in connection with the plan of utilizing the power of Niagara Falls, that some of our most thriving manufacturing centres are not located on water courses, and depend upon coal and not upon water power. The cities of Fall River and New Bedford in Massachusetts are rivalling Lawrence and Lowell in the number of their spindles. Both can obtain their supply of coal by cheap transportation in vessels. The cost of regulating water power is a large item in the use of it, and steam is found to be more reliable and cheaper in many instances. There is no doubt that it would be more economical to generate 5,000 horse power in Chicago by means of coal than to attempt to transmit it from Niagara Falls. A great change, however, in our coal supply would speedily turn attention to the immense waste of energy which is going on at Niagara Falls, and might convert New York State into a beehive of industries.

CHAPTER XIII.

SELF-INDUCTION.

WITH the use of alternating or periodic currents of electricity the phenomenon of self-induction becomes of great importance, and it is desirable to obtain a clear conception of it. The quantity we call self-induction of a wire or coil is generally small compared with its resistance. The larger the self-induction, the more slowly will a continuous electric current rise to its stationary value.

The self-induction is measured by the number of lines of force established by the current. To send out these lines of force requires an expenditure of energy which is measured by the product of the lines of force due to a unit current multiplied by the square of the current, and this multiplied by one half. This energy produces a stress in the medium surrounding the wire or coil. The greater the self-induction, the longer the time it takes to bring a current up to a certain value. When this energy which is stored in the medium is allowed to return to a wire which is coiled around a bar of iron, the magnet thus formed can be demagnetized. This can be accomplished by suddenly placing a wire of small resistance across the poles of the battery which is exciting the electro-magnet. The lines of force are thus withdrawn from the field.

If we have a large amount of self-induction in the field of an electro-magnet—in other words, if we must generate a large number of lines of force to bring the medium around the magnet up to a high degree of stress—we must do a large amount of work. In the case of the electro-magnets now daily employed in commercial operations this energy is measured in tons lifted through several feet.

An idea can be obtained of the energy manifested in the phenomenon of self-induction by observing the great spark and flash of light which manifests itself when the trolley leaves the overhead wire or when the trolley wire breaks. This is caused by the magnetic energy which was stored up in the medium around the wire and motor of the car rushing back from the medium into the wire and decaying there in the form of heat. The heat developed is sufficient to melt the surface of the copper wire constituting the trolley wire. It would take several horse power applied directly for the same instant of time to produce the same heating effect.

When we consider the phenomenon of self-induction we perceive that the early attempts to obtain the velocity of electricity were nugatory, for different values of what appeared to be the velocity of propagation could be obtained with different lengths of the electric circuit. On account of the time that is necessary to establish the strain in the medium along a telegraph wire containing the electro-magnets used in sending signals, the maximum value of the current which actuates the recording apparatus does not arrive until a certain period has elapsed after the sending-key is touched. This period of time depends upon the self-induction or inductance of the circuit; different values of what seems to be the velocity of electricity will therefore be obtained with every circuit.

The early workers in the subject of electricity confined their attention to what they called electricity as it manifested itself on conductors. The idea of a strain in the medium surrounding the conductors, or of energy being stored up in this medium, did not take firm hold of men's minds until Maxwell stated his great hypothesis of the electro-magnetic origin of light. Meanwhile the transformations of energy manifested in the ever-increasing phenomena of electricity demanded the hypothesis of a medium.

The electric energy which propels an electric car resides in the medium around the trolley wire, and the rate of decay of this energy along the wire is what we call the electric current.

The energy of the dynamo, thus, is not transmitted along the wire; it is manifested in the ether of space and in the insulating media about the wire. Lodge, in his Modern Views of Electricity, remarks: "The energy of a dynamo does not, therefore, travel to a distant motor through the wires, but through the air. The energy of an Atlantic cable battery does not travel to America through the wire strands, but through the insulating sheath. This is a singular and apparently paradoxical view, yet it is well founded."

On Poynting's hypothesis, an electrical current should start first at the boundary of the wire near the insulating substance. To-and-fro currents, like those from a discharge of lightning, should be confined to the outside of the conductor. This is found to be true. With such currents, therefore, a hollow conductor is better than a solid one. With an extremely rapid rate of oscillation, far into the billions per second, the electrical current leaves the wire, and may be said to travel through the medium. It then manifests itself as light, and the copper is opaque to light. In speaking of this skin action, Lodge remarks that this action is greatly broken up by making the conductor of bundles of wire, in order to afford the medium surrounding the wire access to every part of the conductor. It is well also to make the conductor expose a large surface to the dielectric.

"A lightning conductor, therefore, should not be a round rod, but a flat strip, or a strand of wires, with the strands as well separated as convenient. I might go on to say here that iron makes an enormously worse conductor than copper for rapidly alternating currents; so it does for currents which alternate with moderate rapidity—a few hundred or thousand a second—like those from an alternating dynamo or a telephone; but, singularly enough, when the rapidity of oscillation is immensely high, as it is in Leyden-jar discharges and lightning, iron is every bit as good as copper, because the current keeps to the extreme outer layer of the conductor, and so practically does not find out what it is made of." *

I have examined the phenomenon of electrical oscillations on iron wire, and I find that Prof. Lodge's conclusions are subject to certain limitations. Oscillations as rapid as ten million a second can magnetize iron; and the period of the waves on iron is also affected by the magnetic nature of the iron.

I have employed in my lectures the following method of illustrating the effect of self-induction: The spark terminals of a Ruhmkorff coil (Fig. 22) are placed

---

* Lodge, Modern Views of Electricity, p. 101.

in a vessel from which the air can be exhausted. This vessel is then connected with an air pump. In the primary circuit of the Ruhmkorff coil is placed a coil of small resistance. When the exhaustion of the vessel is pushed to a certain amount — about fourteen inches of mercury pressure in the vessel—the insertion of a bundle of iron wires in the coil in the primary circuit completely stops the formation of the spark. Work is done in establishing lines of force in the medium about the coil, and less work can be done in producing the spark. If, now, the exhaustion is pushed farther, the white crackling spark is replaced by a reddish-purple glow, and now the increase of self-induction produced by inserting the bundle of iron wires no longer produces any visible effect on the spark. The resistance of the rarefied air has become less, and the change in the self-induction of the circuit is not sufficient to modify the transformation of energy in the transformer. The phenomenon of self-induction can be illustrated in a striking manner by placing a coil of a number of turns but of comparatively small resistance to-

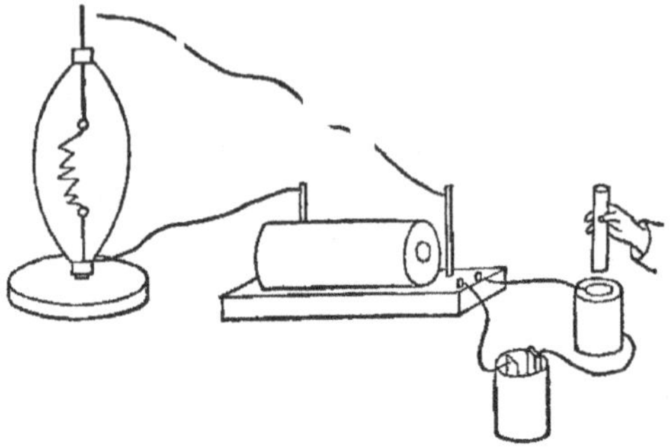

FIG. 22.

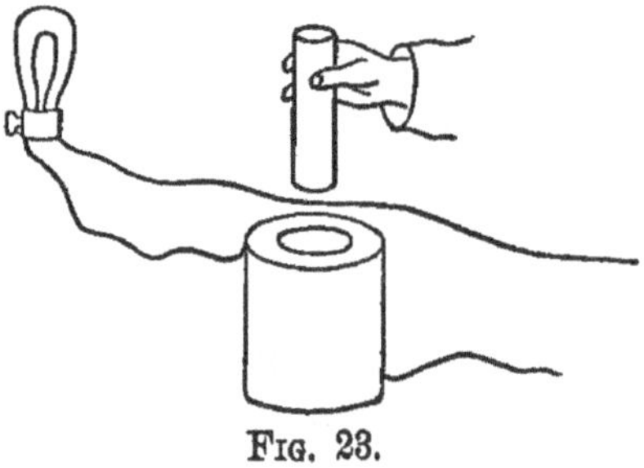

FIG. 23.

gether with an electric light on an alternating-current circuit. When an iron core is thrust into the coil the electric light (Fig. 23) is greatly changed in brilliancy. This change is due to the increased self-induction of the circuit. The change in the flow of induction through the coil has been greatly increased, and this change gives place to an electro-motive force in the wire of the coil which is opposed to the electro-motive force of the alternating-current machine which is feeding the circuit.

Work is required to establish the lines of flow of induction through the iron, and also to withdraw them. The phenomenon of self-induction bears greatly upon the question of lightning rods and the protection of buildings from lightning. To illustrate this, suppose that we should discharge a Leyden jar (Fig. 24) by means of a wire bent into a long loop and provided with two points, A and B, which can be brought near together. It will be found that when the spark occurs between C and D that a spark will also occur between A and B. The electrical current prefers to pass between A and B, and thus to avoid the work of putting lines of force around the loop M. The reason that lightning takes the shortest path is not that this path has the least resistance, but that it has the least inductance. The resistance of the air between A and B is far more than that of the copper loop M. We can therefore conceive of a building being shattered by lightning which is provided with lightning rods, if it requires less work to pass through the building to the

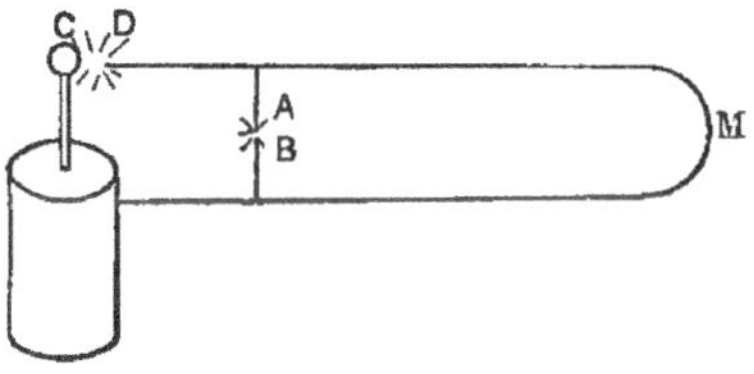

FIG. 24.

ground than to overcome the self-induction of the lightning rod. When we read Franklin's observations on the utility of lightning rods, we perceive that he had no conception of the phenomenon of self-induction. To his mind the electrical fluid took the path of the good conductor in all cases. It may have been that sceptical ones of his day treasured up instances of lightning leaving good conductors in a most unaccountable manner, and were believers in an opposing effect, which we now call inductance.

Prof. John Winthrop wrote to Franklin as follows:

CAMBRIDGE, *January 6, 1768.*

"I have read in the Philosophical Transactions the account of the effects of lightning on St. Bride's steeple. It is amazing to me that, after the full demonstration you had given of the identity of lightning and electricity and the power of metalline conductors, they should ever think of repairing that steeple without such conductors. How astonishing is the force of prejudice, even in an age of so much knowledge and free inquiry!" *

* Sparks, Works of Benjamin Franklin, vol. v, p. 419.

CHAPTER XIV.

THE LEYDEN JAR.

In our further study of the transformation of energy self-induction plays a most important rôle. The coefficient of induction multiplied by one half of the square of the current represents the energy which is stored up in the medium surrounding the wire carrying the current. Any change in the value of the current produces a change in the amount of this energy. Another factor which is never absent on electrical circuits is what is called capacity. A Leyden jar has capacity. We ordinarily say that a certain amount of electricity can be stored up in the jar. The larger the jar, the greater the capacity. The Atlantic cable has a large capacity. It is a long, cylindrical Leyden jar; the conducting wire forms the inner coating and the water the outer coating. In the case of a thunderstorm, the upper layer of clouds forms one coating of a condenser and the surface of the earth the other, while the air between takes the place of the glass in an ordinary Leyden jar or of the gutta-percha of the Atlantic cable. The telegraph wires strung on poles also possess capacity. The wires form one surface of the condenser and the ground the other.

The insulator between two charged surfaces of metal is called the dielectric, and modern inquiry is largely directed to the study of what goes on in the dielectrics under rapidly alternating charges on the metals. There is no doubt that a stress occurs in the dielectric under heavy charges, for the glass walls of Leyden jars are often broken, and we see how the air is cracked, so to speak, by discharges of lightning. The ordinary Edison lamp which is used to light our houses, consisting of a glass bulb inclosing a carbon filament, has considerable capacity. It is a small Leyden jar. If one holds the bulb in one's hand and presents the brass base of the lamp to the conductor of an electrical machine, after a moment of charging one can obtain a shock by touching the brass base of the lamp with one hand while the bulb is held in the other.

When we survey the path we have followed in studying the electric current and the various transformations of energy which are manifested by the rate of change of electro-magnetic induction, we perceive that our attention has been directed mainly to the electrical manifestations on closed metallic circuits. Indeed, to the ordinary mind a wire seems to be the essential part of an electric circuit. Thus, when we discharge a Leyden jar by connecting the outer coating to the inner coating by a wire, we are apt to fix our minds upon this wire as the seat of a momentary electric current, the energy of which is manifested by the spark which results when the jar is discharged. We know that there is a current in the wire when the jar is discharged, for it will melt a wire and decompose water.

Faraday, in 1832, made experiments on the quantity of electricity yielded by the discharge of a Leyden jar; and stated his results as follows: "The decomposition of a single grain of water requires 800,000 discharges of a large Leyden battery. This, if concentrated in a single discharge, would be equal to a very great flash of lightning, while the chemical action of a single grain of water on four grains of zinc would yield electricity equal in quantity to a powerful thunderstorm." *

The prevailing impression is that more electricity can be obtained from a percussion cap filled with moist salt sand in which a piece of zinc wire is immersed, not touching the copper of the cap, than from a discharge of lightning. One can send a signal across the Atlantic cable with such a minute battery; but it is said one can not do this with a spark from a Leyden jar. This last assertion, however, is a mistake. It can be done by means of the spark from a Leyden jar. All that is necessary is to properly transform this spark in the following manner:

Coat any large thin glass vessel on the outside with tin foil and fill the vessel with water. Now connect the outside tin-foil surface with one conductor of an electrical machine and the water inside the vessel with the other conductor of the machine. After a few turns of the machine the Leyden jar becomes charged; and if the water on the inside is connected by means of a wire with the tin foil on the outside, a spark passes when the end of the wire is brought near the tin foil.

Instead, however, of allowing the spark to dissipate itself in light and noise, let us connect the tin foil with one end of a bobbin of well-insulated fine wire of a thousand feet or more in length, but wound compactly on a hollow bobbin. In the centre of this bobbin, entirely disconnected and insulated from the fine-wire bobbin, we will place another coil of coarse wire five or six feet in length wound once around a bundle of iron

---

* Dr. Bence Jones, Life of Faraday.

wire. Across the ends of this coarse wire we will place a small incandescent lamp of from five to six candle power. Now, if the other end of the fine-wire bobbin is brought near the inside of the Leyden jar, a spark jumps and is dissipated through the thousand feet of the fine wire. The little lamp connecting the ends of the coarse coil lights up for an instant. If, instead of the lamp, two platinum wires are placed in acidulated water, and are connected with the ends of the coarse coil, a quantity of bubbles of oxygen and hydrogen gas is given off from each of the platinum wires. The water is decomposed, just as it is by two or three strong chemical or voltaic cells.

We see, therefore, that it is merely a question of transformation. An electric spark can do all that a battery or a dynamo can do. It works, however, for a very short interval of time. It has the characteristic of brilliancy but not of persistence. A simple calculation will enable us to form an idea of the horse power in a spark from a Leyden jar of about a gallon capacity, the glass of which is about one sixteenth of an inch thick and which is charged so that it will give a spark of about two inches long.

Such a spark discharged through our bobbin containing about a thousand feet of wire will light up brilliantly a six-candle-power lamp connected with the coarse-wire bobbin which occupies the centre of the fine-wire bobbin. Now we know from accurate experiments that the spark lasts a few hundred thousandths of a second—it may be three hundred thousandths. We know also a horse power would light from thirty to forty of our little lamps. If there were no loss in transforming the spark, the spark would be equal to one thirtieth of a horse power acting for three hundred thousandths of a second; but there is a loss in transformation of nearly fifty per cent, so the horse power in our spark is twice what we have supposed—two thirtieths or one fifteenth of a horse power.

In a subsequent paper I shall show that we have reasons for believing that the energy of an electric spark an inch in length amounts to thirty or forty horse power; for waves are sent out in the ether in all directions, and these waves are of great energy. At present, however, we are concerned only with a direct transformation of an electric spark into horse power which can be directly measured.

Now, if a spark two inches in length from a gallon Leyden jar is equivalent to one fifteenth of a horse power, what must be the horse power in discharges of lightning which are many hundred feet in length? We have all heard the bells of telephone apparatus ring violently at each discharge of lightning, and, on timing the interval between the flash and the thunder, we find, knowing that sound travels about a thousand feet per second, that the discharge must have occurred a mile away. We should find, on making the necessary calculation, that it would take some hundreds of horse power to produce this electrical effect from the distance of a mile.

Incandescent lamps also often blink at each discharge of lightning from a storm centre at least a mile away. When the discharges occur within a thousand feet the lamps may be nearly extinguished for an instant; therefore the lightning, even at a distance of a thousand feet, holds in check it may be a thirty-horse-power steam-engine which is turning the dynamo machine and supplying the lights. I have no doubt that a discharge of lightning five hundred feet long, if properly directed and controlled, could light for an instant a thousand Edison lights.

It seems at times as if the bolts of lightning grow envious of the great webs of wire which have been spread over our cities, and delight to exhibit their horse power by entering upon electric-light circuits, showing the dynamos how to burn out wires and set fire to buildings. Indeed, one of the most serious concerns of the practical electrician is to devise methods of preventing lightning from breaking and entering. There is a popular superstition that the multiplication of electric circuits in our cities and towns has driven off thunderstorms, but there is no proof that such is the case. The lightning is still an unwelcome visitor, and comes at the most unexpected times. I have no doubt, however, that the multiplication of wires is to a certain extent a safeguard against the exhibition of the horse power of lightning by its destruction of chimney tops, rending of trees, and even the killing of human beings; for the multitude of wires distributes the electrical charge, and it finds a quick passage to the ground in many directions.

I have already said that the practice of combining gas fixtures and electric-light circuits so that one can use gas or electricity is fraught with some danger. The electric wires are often led along the gas pipes, and if lightning should succeed in following the electric wires into the house, it would naturally jump to the gas pipes and seek the ground. If there should be a gas leak, even from a minute pin hole, it might be lighted. If the lightning does not enter the house by the wires, it is possible that a heavy discharge may cause sparks between the electric wires and the gas pipes by induction. I knew of an instance where a spark between the electric-light wires and the gas pipes ignited the gas which streamed from a minute hole in the pipe. If the jet had not been noticed in time the building would surely have been set on fire. The building was provided with lightning fuses; nevertheless, the minute sparks were caused by a lightning discharge. In building a new house one should be careful to keep the electric-wire circuit away from the gas pipes. The practical electrician and the theoretical plumber would doubtless call this a scientific man's superstition, but a long study of electrical sparks has made me respect their wide and varied manifestations of energy.

The more that one knows about the horse power in lightning the more one wonders at the temerity of Benjamin Franklin in drawing lightning from the clouds. He evidently regarded it as a lambent ethereal flame, capable, it is true, of giving disagreeable shocks; he knew its power in rending trees and in setting fire to buildings, yet he could not have had a realizing sense of its horse power. No one to-day would be willing to repeat Franklin's experiment in the manner that he performed it. One could, it is true, lead the wet string into a lake or pond and hold the string with rubber gloves. Most of us, even so, would prefer to be interested spectators rather than participants in the experiments.

Looked at from another point of view, it will be seen that the force required to rend the air—to bore a hole, so to speak, through it as lightning does, to crack it as if it were a piece of glass—must be enormous.

Nothing, to my mind, so strongly illustrates the difference in the intellectual standpoint of the ancients and that of the moderns in regard to science as such a discussion as this upon the horse power of lightning.

Philosophers to-day do not speculate about the primal sources of lightning, but set themselves to work to study the transformations of electricity, with a large hope that they can greatly increase our knowledge of such transformations and with very little hope that they can ascertain what electricity really is. Fancy Faraday's delight, could he have seen the working of the modern transformer, the fine-wire bobbin inclosing the coarse coil with its bundle of iron wires! imagine the immense field of the practical applications of electricity which would immediately have opened to his vision! Cities are now lighted by its means, and it is proposed to transmit to great distances by the transformer the power of Niagara Falls.

Seeing thus the possibility of transforming the electric spark into working horse power, so to speak, a question more or less curious intrudes itself upon one's mind. Is it not possible to make a practical use of the electrical machines which have since the time of Benjamin Franklin played their part upon the lecture table of professors of physics? Unmanageable servants they are often, inopportunely festive, brilliant; but, on the whole, not to be depended upon in all weathers. The professor's Ruhmkorff coil has taken its place in practical life as a transformer such as we have described in this article, but not to transform lightning. It is used in every telephone transmitter in the land, and is employed, as we have said, in lighting cities and in transmitting power long distances. The dynamo machine also has sprung into giant shape from its lecture-room models. Why should not the electrical machine also have its practical development, since we have seen that its sparks can be transformed into horse power?

It apparently requires very little power to turn the disks of the Holtz machine to produce sparks which when transformed will light up for an instant an eight- or ten-candle-power lamp. Why could not one arrange a number of electrical machines in such a manner that, turned by a common shaft, they might charge and discharge Leyden jars continuously, and thus, by means of the transformer we have described, produce light? It is, indeed, conceivable that a great number of large electrical machines of the late improved types could be driven by steam power or water power, and their charges so accumulated in suitable Leyden jars or condensers that a large building could be illuminated. Since it requires time to charge Leyden jars to their full capacity, a great number of large electrical machines would be required, and the intervals between successive discharges of the jars could not be less than one sixteenth of a second, in order that the instantaneous lighting of the lamps should remain on the eye and seem to be continuous.

This endeavour to imitate the action of the ordinary dynamo machine by coupling electrical machines together is much more difficult at present than to procecd in the inverse order and to imitate the action of the electrical machine by means of the dynamo machine. If we send a powerful dynamo current to and fro through the coarse coil of the transformer which we have used to exhibit the horse power of an electric spark, we can readily obtain sparks of several feet in length from the ends of the fine wire of the outer bobbin. By a suitable transformation in the coil we can cause an exhausted globe to become luminous by pointing the finger at it; we can make a lamp glow without leading wires when it is placed anywhere between the walls of a room which are connected with the ends of our transformer. The entire room can be filled with lines of electric force, but it will be at the expense of a large amount of horse power.

To make one lamp glow without leading wires when it is placed anywhere in a small room, requires at present the expenditure of at least twenty-five to thirty horse power. This does not seem to be the light of the future. A large-sized elèctrical machine can even now compete with the dynamo machine in experiments of this sort in producing phosphorescent glow lamps. The dynamo machine, it is true, can imitate all the effects of an electrical machine—its long spark, its phosphorescent glow lamps—but it does this with a great expenditure of horse power. On the other hand, if the electrical machine should endeavour to perform the work of a dynamo machine in lighting incandescent lamps, it would be also at a great expense of horse power.

The practical electrician of to-day has comparatively little use for the Leyden jar, or for the modifications which are called condensers, except in the case of submarine telegraphy. The cable itself, we have said, is a great cylindrical Leyden jar. We can not telegraph along the central core of this cable by the same means which we employ on land lines—that is, by connecting a battery direct to this core and using electro-magnets or Morse sounders. The cable charges under the electro-motive force of the battery, just as a Leyden jar charges under the difference of potential of an electrical machine. The line therefore becomes clogged, so to speak, with the charges sent into it by the different impulses from the battery, and the result is a confusion of signals. The capacity of the cable is so great that time is necessary both to charge and to discharge it. This confusion is made very evident if we attempt to use a telephone and speak over the cable. At a distance of about fifty miles nothing but a murmur is heard; the characteristics of the human voice are completely obliterated by the capacity of the cable. In order to speak under the ocean, some way must be found of neutralizing this capacity. To the electrician whose work lies in the field of telephony the subject of the Leyden jar therefore is of the utmost importance. Meanwhile, to obviate the disturbing effect of the capacity of the cable we do not connect our battery direct to the cable, but, having first charged a condenser by connecting its two coatings to a battery, we discharge the condenser into the cable and use very delicate instruments to record the feeble current which thus traverses the cable. The process of submarine telegraphy is thus simply one of discharging a Leyden jar through it. A comparatively large electro-motive force can thus be used without the danger of heating the wire of the cable and without giving the cable too great a charge.

In the submarine cable, therefore, the practical electrician sees the importance of the study of the Leyden jar. To the scientific man its manifestations have become of greater importance than the electrical effects we have hitherto studied in metallic circuits, for he is persuaded that by the study of the electrical phenomena in the dielectric—the glass, for instance, of a Leyden jar—we shall more clearly understand the relationship between light, heat, and electricity. In an ordinary voltaic cell we can trace the complete circuit of electrical phenomena. The electro-motive force arises in some mysterious way from the two metals of the battery; a current results from this force and the chemical actions in the cell. This current can be traced on the wire connecting the two metals of the cell and in the liquid of the cell. There is no part of this circuit where there is no evidence of an electric current. It is present in the outer wire which rings our house bell; it is evident in the liquid of the cell. In the case, however, of a Leyden jar the current manifests itself apparently only in the wire connecting the outer and inner coating of the jar. The glass that separates these coatings is called an insulator; it does not permit a current of electricity to pass through it. How, then, is the electric circuit completed? Are there no electrical effects in the glass? The great theory of Maxwell—the electro-magnetic theory of light—assumes that there is an electrical effect in the dielectric in the shape of what are called displacement currents, which arise when the electrical energy which is stored up in the ether near the metallic coatings of the jar is made to fluctuate in amount. The cycle of electrical effects manifested by a Leyden jar is made up, therefore, of currents of conduction along the wires connecting the coatings and of displacement currents in the glass. The study of these displacement currents, both from a mathematical and an experimental point of view, is now of the utmost theoretical importance.

The phenomena of the spark from an electrical machine or a charged Leyden jar differ from those exhibited by the voltaic arc mainly in this: the spark studied by Franklin is not one spark, but each spark is made up of a number of sparks which oscillate to and fro. This was first shown by Joseph Henry, and it is a remarkable fact, the importance of which will be seen when we study electrical waves.

When we obtain a great difference of potential between the carbon terminals—for instance, when a spark from a Leyden jar jumps between them—we perceive a marked difference between these poles. A good way to study these differences is to distribute light powder, like lycopodium, on glass, bring wires from a Ruhmkorff coil to opposite points on the glass, and break the circuit of the primary. The light dust is disturbed, and a figure will result on the glass due to the state of electrification of the point of the wire. These figures are different at the positive and negative terminals of the wire, and they have been carefully studied by Bezold and other physicists. Another interesting way to study these figures is to replace the glass plate and the dust by an ordinary photographic plate. On developing the plate a photograph is obtained which shows that the discharge at either the negative or the positive pole is a complex figure made up of both the discharge peculiar to the negative pole and that peculiar to the positive pole. Now it may be asked, How do we know that the figure is a combination of these effects? What we wish to show is, that the discharge from a Leyden jar is a to-and-fro discharge—that it is oscillatory, in other words. In order to prove this, it is necessary to use a revolving mirror to study the discharge. The following is the modified form of Feddersen's apparatus which I have used in my researches, and which I shall often have occasion to refer to in what follows: A little concave mirror is mounted on the armature shaft of a little electric motor, E (Fig. 25). The spark gap is placed just above a sensitive plate, P, which is shielded from the direct light of the spark. If the mirror, M, is at rest, the photograph obtained by reflection at P is simply a zigzag line. When, however, the mirror revolves very swiftly the photograph of the spark is drawn out into a band like a comet's tail, which is seen to be made up of a series of dots of alternate degrees of brightness (Plate I, frontispiece). The darkest dots represent the stronger discharge at the positive terminal, and the light dots the discharge at the negative terminal. The discharge oscillates to and fro until it dies out. What seems to the eye, therefore, as one spark is made of a number. The number depends upon the length of wire in the circuit between the two coatings of the jar and the thickness and size of the jar, or, in other words, what is termed the capacity of the jar. It may be thought that a rough manner of showing the oscillatory nature of the discharge of a Leyden jar would be to perforate a piece of writing paper by the discharge; for, on examining the hole made by the spark, it will be seen to have a burr on both sides of the paper, as if the discharge had passed to and fro through it. This method, however, is an erroneous one, for if the discharge is made nonoscillatory by discharging through a wet string or a suitable liquid resistance, the burrs are also obtained. The phenomenon is therefore due to a species of explosion in the paper, and is not due to the oscillations of the spark discharge.

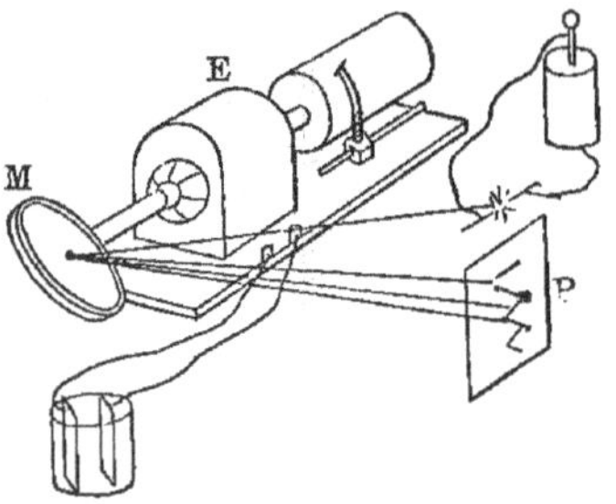

FIG. 25.

If we analyze an ordinary spark by the revolving mirror we discover that it is generally oscillatory. When, therefore, people imagine that they can tell which way a lightning discharge passes, whether from the clouds to the earth, or from the earth to the sky, they must reflect upon this oscillatory phenomenon, and also consider that the interval between such oscillations is less than one ten millionth of a second. An impression remains on the retina about one sixteenth of a second, and the human eye, therefore, can not distinguish direction in the electric spark.

I have used the following method of studying the phenomena at the poles of discharge: The terminals, between which the spark jumps, consist of two thermal junctions. Immediately after the discharge occurs the circuit between the junctions is completed through a galvanometer by a peculiar key. In the case of the oscillatory discharge, the two terminals are heated equally, and there is no movement of the galvanometer needle. When, however, the revolving mirror shows that the discharge is nonoscillatory—and this can be accomplished by putting in a suitable liquid resistance in the path between the two coatings of the Leyden jar—the galvanometer shows that one junction at the positive terminal is more heated than that at the negative.

In 1842, Prof. Henry, in speaking of what was called anomalous magnetism, which was observed in the case of needles magnetized by discharges from Leyden jars—these needles often exhibiting a magnetic condition opposite to that which should result from a current in a definite direction—says:

"This anomaly, which has remained so long unexplained, and which at first sight appears at variance with all our theoretical ideas of the connection of electricity and magnetism, was, after considerable study, satisfactorily referred by the author to an action of the discharge of the Leyden jar which had never before been recognised. The discharge, whatever may be its nature, is not correctly represented (employing for simplicity the theory of Franklin) by the single transfer of an imponderable fluid from one side of the jar to the other; the phenomenon requires us to admit the existence of a principal discharge in one direction, and then several reflex actions backward and forward, each more feeble than the preceding, until the equilibrium is obtained. All the facts are shown to be in accordance with this hypothesis, and a ready explanation is afforded by it of a number of phenomena which are to be found in the older works on electricity, but which have until this time remained unexplained." *

* Scientific Writings of Joseph Henry, vol. i, p. 201, Smithsonian Institution, Washington.

CHAPTER XV.

STEP-UP TRANSFORMERS.

THE range of transformations of energy which the Ruhmkorff coil exhibits is by no means exhausted. When we succeed in producing from a battery a spark similar to that generated by an electrical machine, we have, by the use of a fine-wire coil wound upon a coarse-wire coil, exalted the electro-motive force of three or four voltaic cells—it may be eight volts—to perhaps twenty to thirty thousand. Starting from this great difference of potential, is it possible to treat the Ruhmkorff coil as a battery, and to still further exalt its difference of potential? This has been done by Prof. Elihu Thomson, and also by Tesla, and their experiments are most brilliant in the subject of the transformations of energy.

Prof. Thomson has succeeded in producing sparks five feet long, and states his belief that sparks twenty feet in length could be obtained by an extension of the method which he employed. This method was as follows: An open single-layer coil of coarse wire with about ten turns constituted the new primary. Upon this was wound about three hundred turns of fine wire. This latter coil constituted the new secondary. Both coils were immersed in oil. On passing through this new primary coil sparks from a Ruhmkorff coil very long sparks can be obtained from the secondary. This arrangement constitutes a species of double Ruhmkorff coil, or two step-up transformers. Instead of using a battery to excite the first Ruhmkorff coil, an alternating-current dynamo is employed. Leyden jars are connected to the terminals of the first Ruhmkorff coil, and these are rapidly charged by the dynamo. A spark gap is interposed to the circuit between the first Ruhmkorff and the second, and the resulting spark is blown out by a jet of air under high pressure. The second Ruhmkorff gives sparks of the extraordinary length of five feet or more. This length of spark far exceeds that given by the most powerful electrical machine. If a plate of glass is interposed between the terminals of the second coil, its surface is covered with brilliant ramifications of violet-coloured sparks.

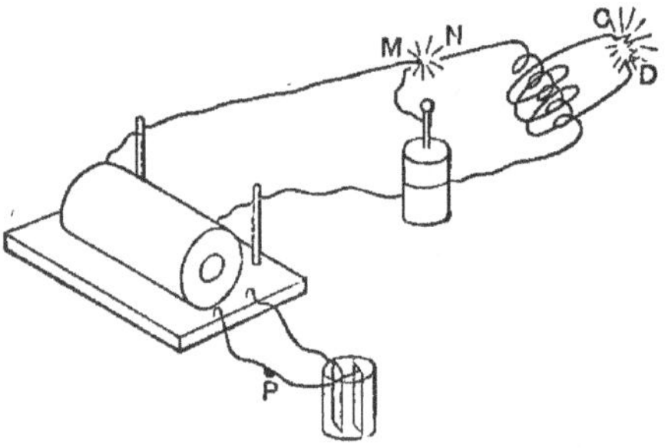

FIG. 26.

To heighten the results, as above mentioned, the spark produced in the secondary of the first Ruhmkorff at M N (Fig. 26) is constantly blown out by a strong blast of air. This blast serves the function of the break in the primary at P. A very high electromotive force can thus be obtained between C and D.

I have said that we have here two step-up transformers in the shape of two Ruhmkorff coils. A Leyden jar is interposed, as shown in the figure, in order to increase the quantity of electricity which is discharged through the primary of the second transformer, and also to act as a second alternating machine.

With this arrangement of the apparatus very strong insulation is needed, for the entire line is charged with electricity of high tension. The leading wires are luminous in the dark, from the brushlike discharges which are given off in every little break in the insulation of the wires. It must be remembered that the lines of force endeavour to leave the positive terminal of the Ruhmkorff coil at any point which offers a shorter passage to the negative terminal. In one sense, therefore, one should not be astonished to find that an exhausted globe will become luminous when it is attached to merely one terminal of the Ruhmkorff coil, for the walls of the room and the floor may constitute the other terminal of the coil, and the lines of force, stretching out and pervading the space in the room, converge on matter which affords in any way the easiest passage. Thus the forefinger glows when presented to either terminal of the coil. The lines of force find on the human body this short passage. When the electromotive force or electrical intensity is greatly enhanced the tendency of the lines of force to manifest themselves through the space inclosed in any ordinary room is greatly increased. At the same time the to-and-fro currents or electrical oscillations on the leading wires tend to confine themselves to the surface of these wires. This can be shown in a popular manner by connecting the terminals C and D (Fig. 26) by a thick copper rod, and, holding one terminal of an ordinary incandescent lamp in one hand (Fig. 27), touch the copper loop with the other terminal and also grasp the loop with the other hand. Only a slight shock is felt, and the currents passing over the surface of the human body raise the carbon filament of the lamp to a brilliant incandescence. The surface of the body is greater than that of the copper loop; and the to-and-fro currents are not compelled to confine themselves to a thin film of copper constituting the surface of the copper. When steady currents are passed through the loop, they are not confined to the outer layer of the copper, and find an easier passage through the section of the copper loop than through the human body. With steady currents it is impossible to light the lamp in the above

FIG. 27.

manner through the human body. The effect of increasing the frequency of to-and-fro currents of electricity is thus to drive them to the surface of metallic conductors. When the frequency or rapidity of vibration is enormous, a rod of copper may not afford any better passage than a rod of glass. Hertz, in one of his papers, points out that our present nomenclature is limited, and only applies to the special cases of steady currents. With enormously rapid to-and-fro currents, a piece of copper acts like an insulator and prevents any to-and-fro currents from passing through it, whereas a piece of glass transmits them unimpaired. A thick disk of copper properly placed between the two coils of a step-up transformer can completely stop the electrical oscillations from reaching a lamp connected with the secondary coil of the transformer, whereas a plate of glass allows them to pass on unimpeded. Glass, therefore, is a better conductor for electrical oscillations of high frequency than copper. We are thus approaching the behaviour of light to these two substances.

Is it not possible, therefore, by enormously increasing the frequency of electrical oscillations, to drive them completely off metallic conductors and compel them to be propagated through the ether of space? If we could do this, and if our oscillation should meet metallic conductors, their energy would decay or be absorbed in such conductors, just as light waves are absorbed by nonconductors of light.

The latter supposition leads us again to Prof. Poynting's view of the decay of electro-magnetic radiations which, proceeding from the sun, pervade all space about us. This decay produces the phenomenon of the electric current.

With a sufficiently powerful electro-motive force we can produce such a stress in the medium in an ordinary room, or, in other words, we can polarize the medium and make it such a storehouse of electric energy, that we can light a little electric lamp anywhere in the room without wires. Carrying this lamp in our hand, it will light up when we enter the room and be extinguished when we leave it. Tesla has shown the possibility of this by making the room a great Leyden jar, of which the walls form the opposite coatings and the air takes the place of the glass. The charge upon the coatings of such a jar, or, in other words, the walls of the room, is made to alternate with great rapidity, and electric waves fill the air, giving periodic to-and-fro movements to it.

The molecules in a little rarefied bulb are thus set in very rapid motion, and by their impact on suitable substances can raise the latter to incandescence, and by their mutual collisions can also fill the rarefied tube with a luminosity.

Tesla, in his remarkable lectures on the effects produced by currents of high frequency, shows that a suitably constructed lamp can be made to glow in any space by being connected merely with one terminal of a transformer. A little motor can also be made to revolve by being attached to one wire; the ordinary electric motor requiring two wires, one connected to the positive pole of the dynamo and the other to the negative pole. In considering these experiments we must remember, however, that the electric circuit in both cases of the light and the motor is completed through the medium from the positive pole of the exciter to the negative pole. In other words, lines of electrostatic force extend through the medium, through the walls of the room, back to the negative pole of the generator. We can not isolate an electric effect at one spot, or consider that our lines of stress stop at this spot. The apparatus by means of which Tesla produced remarkable luminous effects is similar to that of Thomson, and can be characterized as a step-up transformer of the second order. Instead, however, of using an alternating machine of comparatively low rate of alternation, Tesla employed one giving about fifty thousand alternations per second. With the very high potentials obtained in his second step-up transformer he was able to excite the molecules of rarefied gases to a rapid rate of movement, and by their impact on suitable matter to produce vivid light.

Prof. Crookes's experiments on radiant matter in highly exhausted tubes may be said to have first drawn inventors' minds to the possibility of obtaining light by other means than by employing steady currents of electricity to raise matter to incandescence. In one form of the Crookes tubes one terminal of a Ruhmkorff coil ends in a concave mirror (Fig. 28) inside an exhausted globe. At the focus of this globe (*b*) is placed a bit of platinum on a glass stem. The energy streaming from *a* is reflected to the focus *b*, and raises the bit of platinum to incandescence. This lamp may be said to be the forerunner of the later attempts of Tesla to produce light by high electro-motive force. The tube is of importance also in X-ray photography. The commercial employment of strong to-and-fro currents enables one to greatly magnify the results obtained by Crookes.

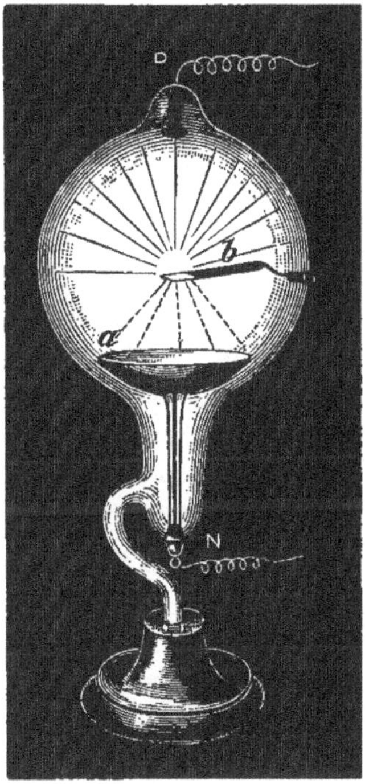

FIG. 28.

The experiments of Tesla to obtain a light with a small amount of expenditure of energy are extremely suggestive. At present, however, in order to produce the luminescence of such economical lamps we are compelled to employ powerful dynamos and transformers. We need an engine of at least ten horse power to produce the conditions for such an economical lamp of a few candle power. The problem is, to produce the conditions economically. I can illustrate this problem by a species of *argumentum ad hominem.* An electric light of feeble power can be produced by shaking a small amount of mercury in a glass tube which has been partially exhausted of air. The friction of the mercury against the glass walls of the tube produces an electrification which in turn leads to a heightened oscillation or movement of the molecules of air still left in the tube. No heat can be detected in this light. We have produced considerable light by electrical excitation with very little heat, but the amount of heat supplied to the human body in the shape of food is very great, and we have to go through an expensive transformation to produce our little light.

It will be well at this stage of our study of transformations of energy accomplished by the inventions of man to examine into the degree of perfection which has been obtained. The efficiency of any engine is the amount of work it can perform compared with the amount of energy given to it in the shape of fuel. "The steam engine has an efficiency of about ten per cent; the efficiency of the best dynamo machines is ninety per cent; therefore only nine per cent of the energy in the coal is transformed into electrical energy. In the conversion of this electrical energy into light about ten per cent is lost in the conductors, and we have cousequently in the lamps only 0·081 of the energy in the coal. Of this energy in the lamp about ninety per cent is expended in producing heat, ten per cent only being useful for the production of light. Thus, the efficiency of the electric lamp is only 0·0081, or about one per cent." *

This result is well calculated to repress any feeling of exultation we may have in contemplating the

* Palaz, Industrial Photometry, p. 265.

great field which has been opened by the transformations of energy accomplished by the dynamo machine. We can congratulate ourselves, however, that the electric light is more efficient than other forms of light. It has been computed that the energy consumed in producing a light of sixteen-candle power by kerosene is 42·86 watts (a watt is $\frac{1}{746}$ of a horse power). An Argand gas burner of twenty-two-candle power consumes 68·8 watts; an incandescent electric light, 3·5 watts per candle; the arc light, 0·8 watt per candle.

We fail, thus, in utilizing the energy in the coal, and when we produce a light we convert most of the small amount of energy we obtain from the coal into nonluminous heat waves. It has been computed that ninety-five per cent of the energy expended in producing a light goes to the production of waves of the ether which do not affect the eye. Lodge remarks that we are in the condition of an organist who, in order to sound certain high notes of his instrument, is compelled to sound all those of the keyboard. Two great practical questions in the transformation of energy thus confront us: How to utilize to a greater degree the energy stored up in the coal, and how to produce a light rich only in those rays which appeal to our senses as light.

Ebert * has constructed an economical lamp in the following manner: A is an exhausted glass globe (Fig. 29) with an inner glass stem, B, on the end of which is a paste of phosphorescent paint—"Grün blaue Leucht farbe." E_1 and E_2 are tin-foil rings, which are glued to the glass globe. These rings are connected to the two terminals of the finer coil of the step-up transformer. Under the action of the to-and-fro currents of the trans-

* Ann. der Physik und Chemie, No. 9, 1894.

former the plates E_1 and E_2 are electrified, and the lines of force fluctuating through the rarefied globe raise the phosphorescent paint to a high degree of luminosity. Ebert calculates that such a lamp consumes from 1,500 to 2,000 times less energy than the amylacetate unit lamp, and that this amount of energy is in the neighbourhood of one millionth of a watt, the watt being $\frac{1}{746}$ of a horse power. In order to avoid the inevitable losses which arise in conveying such high-tension effects any distance, Ebert suggests that a little step-up transformer can be put in the base of each little lamp. The necessity of electrical turning in the experiments of Ebert is perhaps the most interesting fact in regard to the endeavour to obtain an economical lamp by means of to-and-fro high-tension currents. He found that unless the circuit in which the lamp was placed was in resonance with the exciting circuit, the lamp did not light up to its full brilliancy.

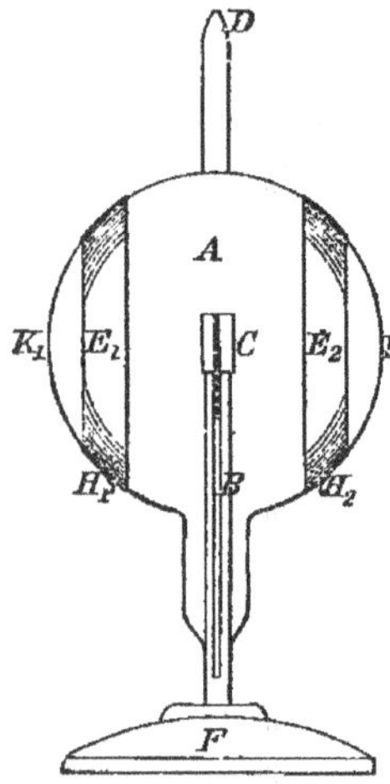

FIG. 29.

In an interesting paper on the cheapest form of light, Prof. Langley calls attention to the fact that there is an enormous waste of energy in the ordinary methods of producing illumination. In the ordinary Argand-burner gas flame this waste for illumination purposes can be shown to be something over ninety-nine per cent of the radiant energy emitted by the lamp. As Prof. Langley points out, "this waste comes from the necessity of expending a large amount of heat in invisible forms, and each increase of light represents not only the small amount of heat directly concerned in the making of the light itself, but a new indirect expenditure in the production of invisible calorific rays. Our eyes recognise heat mainly as it is conveyed in certain rapid ethereal vibrations associated with high temperatures, while we have no usual way of reaching these high temperatures without passing through the intermediate low ones, so that if the vocal production of a short atmospheric vibration were subject to analogous conditions, a high note could never be produced until we had passed through the whole gamut, from discontinuous sounds below the lowest bass, up successively through every lower note of the scale till the desired alto was attained." *

The phenomena of phosphorescence as it is manifested in fireflies, seems to form an exception to this rule. The light emitted by this insect can be produced artificially by raising a body to 2,000° F. No sensible heat, however, accompanies the firefly's light, and indeed this can also be said of the light in Geissler tubes. It is assumed that the firefly's light is produced without the invisible heat that accompanies our usual processes, for the spectrum of the firefly's light falls off more rapidly toward the red end than the spectrum of a candle, for instance. Prof. Langley, in his Memoir, which we have quoted, gives in detail the delicate measurements with his bolometer, by which he obtained the following results with the Cuban firefly (*Pyrophorus noctilucus*), an insect about one inch and a half long and half an inch wide. He concludes that Nature produces this cheapest light with about the four hundredth part of the cost of the energy which is expended in the candle flame.

---

* American Journal of Science, vol. cxl, 1890.

CHAPTER XVI.

LIGHTNING.

By means of the alternating-current dynamo and by the Ruhmkorff coil we have been able to transform the energy of coal into the energy manifested by lightning, for the long sparks of Thomson are lightning discharges. Electro-magnetic waves from the sun produced the coal, and we shall see later that from the coal, and by means of the dynamo and the step-up transformers, we can obtain again electro-magnetic waves. After one hundred and fifty years we have come back to the study of sparks and the behaviour of Leyden jars. We have returned to the domain of science, which Benjamin Franklin may be said to have suddenly illumined when he drew lightning from the clouds.

We return, however, after a long study of the mechanical equivalent of heat and with scientific knowledge greatly increased by exact measurements. We are conscious of the great truth that whenever we can test our electrical theories by heat experiments, following out the line of work indicated by Count Rumford, we are certain to obtain a residuum of truth. The study of the transformations of energy by means of instruments which measure the equivalence in heat of the electrical actions we observe is the final method we must adopt to test any physical theory of electricity.

Before the epoch of Count Rumford the amount of exact experimentation in the subject of physics was extremely small, and it is not to be wondered at that men attributed to mysterious effluvia, to caloric, and to phlogiston the cause of the various transformations of energy which they witnessed. They had no measures of comparison. Benjamin Franklin's experiments have stood the test of time, but his theory of electricity has long since ceased to have value in the scientific world, largely because it had not the weight of quantitative measurements behind it. Let us look at his theory for a moment, and then examine our present conceptions of the lightning flash, which was of such absorbing interest to him.

The following account of Franklin's fluid theory of electricity was presented to the Royal Society in 1851 by William Watson, F. R. S.:

"This ingenious author (Franklin), from a great variety of curious and well-adapted experiments, is of opinion that the electrical matter consists of particles extremely subtile, since it can permeate common matter, even the densest metals, with such ease and freedom as not to receive any perceptible resistance; and that if any one should doubt whether the electrical matter passes through the substance of bodies or only over and along their surfaces, a shock from an electrified large glass jar taken through his own body will probably convince him.

"Electrical matter, according to our author, differs from common matter in this, that the parts of the latter mutually attract, and those of the former mutually repel each other, hence the divergency in a stream of electrified effluvia; but that though the particles of electrical matter do repel each other, they are strongly attracted by all other matter. From these three things, viz., the extreme subtilty of the electrical matter, the mutual repulsion of its parts, and the strong attraction between them and other matter, arises this effect, that when a quantity of electrical matter is applied to a mass of common matter of any bigness or length within our observation (which has not already got its quantity), it is immediately and equally diffused through the whole.

"Thus common matter is a kind of sponge to the electrical fluid; and as a sponge would receive no water if the parts of water were not smaller than the pores of the sponge, and even then but slowly if there was not a mutual attraction between those parts and the parts of the sponge, and would still imbibe it faster if the mutual attraction among the parts of the water did not impede, some force being required to separate them, and faster if, instead of attraction, there were a mutual repulsion among those parts which would act in conjunction with the attraction of the sponge, so is the case between the electrical and common matter. In common matter, indeed, there is generally as much of the electrical as it will contain within its substance; if more is added, it lies without upon the surface and forms what we call an electrical atmosphere, and then the body is said to be electrified."

"Common fire is in all bodies, more or less, as well as electrical fire. Perhaps they may be different modifications of the same element, or they may be different elements. The latter is by some suspected. If they are different things, yet they may and do subsist together in the same body. When electrical fire strikes through a body, it acts upon the common fire in it, and puts that fire in motion, and if there be a sufficient quantity of each kind of fire, the body will be inflamed. When the quantity of common fire in the body is small, the quantity of the electrical fire (or the electrical stroke) should be greater; if the quantity of common fire be great, less electrical fire suffices to produce the effect. . . . Metals are often melted by lightning, though perhaps not from heat in the lightning, nor altogether from agitated fire in the metals. For, as whatever body can insinuate itself between the particles of metals and overcome the attraction by which they cohere (as sundry *menstrua* can) will make the solid become a fluid, as well as fire, yet without heating it, so the electrical fire, or lightning, creating a violent repulsion between the particles of the metal it passes through, the metal is fused." *

Franklin had no measuring instruments or insulated wire at hand, and his entire electrical apparatus consisted practically of merely an electrical machine and Leyden jars. To him the great manifestation of atmospheric electricity in the lightning flash constituted the chief object of study. He bestowed no thought on what is now the great object of our scientific study—the dielectric which is pierced and shattered by the lightning discharge.

Let us return for a moment to the object of Franklin's studies, the lightning discharge, and examine it by our modern methods.

It is well known that when air is subjected to a sudden strain at the moment of an electrical discharge it acts like glass or a similar elastic solid, and is cracked

* Works of Benjamin Franklin. Jared Sparks. Vol. v., p. 221.

in zigzag fissures; indeed, the resemblance between the ramifications of lightning and the seams produced in plates of glass by pressure has been commented upon by various observers. Photographs of powerful electric sparks lead one to conclude that a discharge of lightning makes way for its oscillations by first breaking down the resistance of the air by means of a disruptive pilot spark; through the hole thus made in the air the subsequent surgings or oscillations take place.

I have remarked in a previous chapter that if a powerful discharge from a Leyden jar perforates a piece of cardboard a burr is raised on both sides of the paper. The believers in the two fluid theories of electricity explained this phenomenon by saying that the positive electricity passed in one direction, and the negative in the opposite direction.

In order to study more carefully the cause of this burr on both sides of a sheet of cardboard which is perforated by an electric spark, I arranged two Leyden-jar circuits, one of which was oscillatory and the other was nonoscillatory; and I studied the perforations in a sheet of cardboard produced by the two kinds of discharges, in order to see if the cause of the burr on both sides of the paper was due to the oscillations of the discharge. If this were so, the unidirectional discharge should give a burr only on one side of the paper. In order to charge the jars I employed an alternating dynamo with a step-up transformer, or, in other words, a Ruhmkorff coil. The Leyden jars were connected to the terminals of the Ruhmkorff, and in this manner a torrent of sparks was produced. Photographs of the sparks from the jars were obtained by means of a revolving mirror. Having got a suitable oscillatory discharge, a disk of cardboard was mounted on a revolving shaft, and was perforated by the sparks when in rapid movement. In this way a large number of perforations could be studied. A liquid resistance was then inserted in the circuit of the Leyden jars, and increased until the photograph taken showed that the oscillations had been damped by the resistance, and only a unidirectional discharge remained. The revolving disk of cardboard was then perforated by these unidirectional sparks, and on comparing the two sets of perforations it was found that the burr occurred on both sides of the paper as well with the unidirectional sparks as with the oscillatory sparks. It therefore does not arise from the to-and-fro movement of electricity. I am inclined to consider it as due to an explosion in the paper produced by the heated air. It is well known that gunpowder can not be fired by the spark from a Leyden jar unless one interposes a wet string in the circuit. This wet string acts like the liquid resistance in the above experiments; it damps the oscillation, and diminishes the explosive effect of the heated air, which drives the particles of gunpowder asunder. The phenomenon certainly has no bearing upon the two fluid theory of electricity.

In examining the early photographs, taken by Feddersen, of electric sparks, one perceives that the electric oscillations tend to follow, for at least some hundred thousandths of a second, the path made by the pilot spark; and there are observers who believe that by rapidly moving a camera they have obtained evidence that successive discharges of lightning follow the same path. Prof. Lodge has protested, with reason, against the conclusions drawn by the method of "waggling" the head or a camera; for the movement of the head or the camera certainly requires the hundredth of a second, while the discharge of lightning is over in less than one hundred thousandth of a second.

The method of photographing electrical discharges by means of a revolving mirror seems to be the best method of studying the behaviour of air which is suddenly subjected to the electric strain. I have therefore examined this behaviour with more powerful means than those employed by previous observers; and it may be well to recall here the fact that in lightning discharges high electro-motive force and great quantity are frequently combined in a very short interval of time. The modern alternating machine, therefore, and the device of the transformer enable one to study the character of lightning more successfully than is possible by means of an electrical machine; for both the electro-motive force of a discharge and its quantity can be adjusted over a wide range. In my study of this subject I employed an alternating machine which gave three hundred to four hundred alternations per second, and a current of from fifteen to twenty ampères. By means of a step-up transformer and an oil condenser, discharges of high electro-motive force and great quantity could be readily obtained. The method of the excitation of a Ruhmkorff coil or transformer by means of an alternating dynamo—due originally to Spottiswoode—has placed in the hands of the experimenter, as I have said, powerful means of studying electric discharges; and by the device of an air blast, or other contrivance for obtaining a quick break in the continuity of the electrical discharges, high electro-motive force can be obtained. In certain sparks which I studied, the interval between the oscillations was found to be about one hundred thousandth of a second, and the electrical discharge followed exactly the same path in the air for three hundred thousandths of a second. During this length of time every sinuosity in the pilot spark is exactly reproduced. I employed terminals of tin; and, in one case, a mass of melted and vapourized tin remained suspended in the air for at least three hundred thousandths of a second before it was dissipated in a cometlike tail. During the three hundred thousandths of a second, therefore, the air remained passive while the electrical oscillations took place. During this time it is fair to conclude that the heat produced by the passage of the spark was not sensibly conducted away. If conduction of heat had taken place, the electrical resistance of the air path would have been sensibly altered, and the path of the discharge would have changed in form. Here, I think, we have an interesting limit to the time it takes atmospheric air to respond to the phenomenon of heat conduction.

I have said that the discharges I employed were powerful both in regard to electro-motive force and to quantity. Iron terminals one quarter of an inch in diameter were raised to a white heat by the continuous passage of the sparks, and globules of the melted metal were formed. When the sparks were passed through the secondary of a transformer three fifty-volt Edison lamps placed in multiple in the primary of the transformer, which consisted of merely two layers of thick copper wire, were lighted to full incandescence. The spark from two large glass condensers of 5,000 electrostatic units each, excited by an electrical machine, and passed through the secondary of the same step-down transformer, barely raised a six-volt lamp in the primary to a red heat. The study of the efficiency of step-down transformers in thus transforming transient currents of high potential to transient currents of low potential and comparatively large current, enables one to obtain an estimate of the high potential of lightning, and of the current which accompanies its fall of potential. Thus, if C denote the current in the lightning discharge, and E the electro-motive force, C′ and E′ the corresponding quantities in the circuit of the step-down transformer, A the efficiency of the transformer, we shall have

$$C'E' = ACE.$$

The element of time and the mode of transformation must be considered in any estimate of the amount of energy in a lightning discharge. Although a powerful spark of electricity from two Leyden jars, each of 5,000 electrostatic units, is incapable of decomposing water directly, yet by its passage through the secondary of a step-down transformer it can decompose the water with great evolution of the gases; and it is probable that an ordinary discharge of lightning of a few hundred feet in length, I have before remarked, could light for an instant many thousand incandescent lamps if it were properly transformed by means of a step-down transformer. Indeed, the ringing of electrical bells and the melting of electrical fuses are of common occurrence during thunder-storms, and manifest the energy of lightning discharges. During a recent visit at a summer hotel which was lighted by incandescent lamps, I was much interested to observe that the lamps blinked at every discharge of lightning, although the interval which elapsed between the blinking and the peals of thunder showed that the storm was somewhat remote. This effect was doubtless due to induction produced by the surgings of the lightning discharges; for in heavy and in near discharges the lights were completely extinguished, although no fuses were burned. Electric-light wires and gas pipes should never be contiguous, for no lightning guard or protector can insure that minute sparks, due in some cases to resonance effects, may not arise.

The study of the disruptive or oscillatory discharge of lightning is closely related to that of the brush discharge and the phenomenon of the aurora borealis, for the disruptive discharge ceases to be disruptive after a few hundred thousandths of a second, and partakes of the nature of a brush discharge. The zigzag fissure in the air disappears, and only the spark terminals glow. Recent experimenters have exhibited as a marvel the lighting of a vacuum tube through the human body by grasping one terminal of a suitable transformer with one hand and by holding the vacuum tube in the other hand. It must be remembered, however, that the lines of force proceed from the hand which holds the vacuum tube through the air and the walls or floor of the room to the other terminal of the transformer. We can change this brush discharge or luminosity at either terminal of a transformer into a disruptive discharge by lessening the distance between the terminals or by increasing the electro-motive force.

I am fully aware that the oscillatory discharge of lightning with its disruptive effects, which I have noted, its permanence of path, and the fading of the disruptive discharge into the brush discharge or mere luminosity at either of the spark terminals, is a far simpler phenomenon than the luminosity produced in rarefied tubes; for in the latter phenomenon we have the dissociation and impact of molecules, and we must consider all the problems of atomic motion in addition to those of the oscillatory nature of electrical waves. It is not my purpose to enter into a consideration of the molecular movements involved in oscillatory discharges in vacuum tubes; but having discussed some of the general features of discharges of electricity in air at the ordinary pressure, I shall endeavour to trace the connection between such discharges and the phenomenon of the aurora borealis. To my mind, the luminosity in a vacuum tube, the glass exterior of which is held in one hand while the other hand grasps the terminal of a Ruhmkorff coil, closely represents the phenomenon of the northern lights; for we have in this case a discharge of electricity from a higher level to a lower through a rarefied medium.

I have said that we can pass by insensible gradations from the condition of the brush discharge to that of the disruptive discharge. By intercalating a noninductive water resistance and a vacuum tube between the terminals of a suitable transformer, we can exactly imitate the phenomena observed when the vacuum tube is held in one hand while the other hand grasps one terminal of the transformer. In this case the water resistance takes the place of the resistance of the air of the room. The intensity of the discharge being thus much diminished, one can readily study various manifestations of stratification which may, perhaps, be termed transitory stratifications in distinction to the stationary wavelike forms observed in narrow tubes. The transitory stratifications can be produced at will by touching suitable points of a vacuum tube with the finger or by connecting such points with the ground. Such stratifications are stationary as long as the ground connection is maintained, and are independent of the rate of the alternating machine which excites the transformer. It is evident that the condenser action of the vacuum tube plays an important part in this phenomenon. In observing the striæ and columnar form of the waving of the light excited in this manner in vessels or tubes filled with rarefied gases, one is led to believe that the stratified form of the aurora borealis is produced in a similar manner.

The pulsation, therefore, of the aurora is in no way, I believe, connected with any phenomenon of the oscillatory discharge; yet certain writers have intimated that the glowing of vacuum tubes which are connected with only one terminal of a transformer and the light of the aurora are due to millions of electrical oscillations per second. Now it is impossible to study the question of the rate of oscillation of the brush discharge by means of Fedderson's method, for the light of the discharge is not sufficient to produce a photograph. A brief consideration, however, of the laws of electrical oscillations shows, I think, that such writers are mistaken; for the rate of decay of the amplitude of such oscillations is expressed by the factor $\epsilon^{\frac{Rt}{2L}}$. In the case of the brush discharge, although we may be dealing with very small values of self-induction, L, and small values of time, *t*, we have, on the other hand, great values of R. I believe, therefore, that the brush discharge is reduced to the case of one throb, which is analogous to the pilot spark in disruptive discharges (Plate I, frontispiece). In regard to the aurora, it may be urged that the resistance of the rarefied air is not enormous. In answer to this it can be said that the phenomenon of the aurora can be best reproduced by intercalating a tube of rarefied air with a very large water resistance between the terminals of a suitable transformer. The supposition that the aurora is produced by the action of extremely rapid electrical oscillation on molecules of rarefied air is not borne out by the theory of transient currents; and experiment shows that the phenomenon of the waving and apparent stratification observed at times in the aurora is due to the redistribution of the lines of force which is produced by suitable earths or conductors in the shape of regions of cloud or moisture.

The comparatively small resistance of the electric spark in air, noticed by many observers, is due, I believe, to the permanence of path; for this path is intensely heated, and is practically a charred hole in the air. When this path no longer becomes such a hole and the heated air rises and is dissipated, the oscillations of the electric spark become rapidly damped, and we have the phenomenon of the brush discharge—a glow at each of the spark terminals without a disruptive discharge; the lines of force crowding from one terminal seek the other terminal through the air of the room, and in passing through rarefied air the energy along the lines of force is manifested by molecular actions which are apparently protean in form. I see, therefore, no evidence for believing in the rapid oscillation of the aurora.

The more or less general use of Edison lamps enables one to try many experiments which are possible with vessels from which the air has been partially withdrawn. An incandescent lamp suspended by its brass socket from a conductor of an ordinary electrical machine glows when the machine is excited; the glow does not come from the carbon filament, but from the motion of the molecules of air left in the tube, which are greatly excited by the electrical energy which is being stored up in this little Leyden jar. The whole interior of the globe is filled with a luminescence which seems to be more intense along the carbon filament.

If one takes one of these globes, the filament of which is broken and which still preserves its partial vacuum, one can store up in it sufficient energy to provide luminescent effects for it may be an hour after the lamp is removed from the machine. To accomplish this, one should hold the bulb of the lamp in the hand and charge the broken filament through the brass socket by touching the latter to the conductor of the electrical machine. When the bulb is removed from the machine the light in it disappears. It can be made to reappear, however, at will by simply touching the socket to the table, the floor, or the wall of the room. One can thus provide himself with a little electrical lamp, which he can light by simply touching it to the wall of the room, or to another person's body, or to a piece of tin foil glued to a plane of glass. Its light is feeble, to be sure. It is like a firefly's light—a mysterious radiance. In watching its fluctuations as one's hand is moved over the glass globe one is forcibly reminded of the streaming of the aurora borealis, and one can not but conclude that the wavering light of this phenomenon is due to the same cause—the slow discharge through rarefied air of the electrical charge on the condenser formed by the upper layers of clouds and the lower strata of humid air. Not all incandescent lamps enable one to thus store up luminescence. If the lamp, however, preserves a suitable degree of rarefaction the phenomenon I have described can readily be produced.

With a suitably broken carbon filament an interesting electrostatic effect can be noticed also with Edison glow lamps. Holding the bulb in one hand and bringing the brass socket in contact with the positive prime conductor of an electrical machine, the little globular condenser receives a charge. On touching the table with the brass socket the filament immediately begins to vibrate, and ceases to vibrate when it is held in the air, not touching any object. Immediately on touching the table, the floor, the wall, or any object, the filament begins to vibrate again, causing a ringing note like that of a feeble electric bell. Indeed, it is an electric bell, which can be made to ring for, it may be, an hour after the charging. It is evident that the broken filament either positively or negatively charged is attracted to the oppositely charged glass surface, and, having lost a portion of its charge, its elasticity causes it to swing back, and by its connection with the ground through the table a difference of potential is again established between it and the charged glass surface, and the phenomenon is repeated. If a lamp with unbroken filament is charged by thus holding it in contact with the conductor of the electrical machine and is afterward lighted by an electric current the glowing filament, if it is a thin and flexible one, will continue to vibrate for many minutes.

The cause of atmospheric electricity is not well understood. It is certain, however, that it is not due to the evaporation of water, for exhaustive experiments have never detected the slightest electrification due to evaporation. The friction of particles of water, however, against material substances is abundantly able to produce a high electrification. This is proved by the Armstrong electrical machine, by means of which jets of water spray were forced through nozzles, which became strongly electrified. A curious case of this action lately came to my attention. The operators on a telephone circuit were much troubled by sparks occurring on the line, and it was found that the circuit was electrified by means of a locomotive which, stationed on a side switch, blew off steam against the overhead wires.

The friction of dust particles must also be a potent cause of electrification. The tops of the pyramids in sand storms are strongly charged. Prof. Lodge has made the following suggestive remark: * "It has been discovered by meteorologists that thunderstorms are often associated with curious V-shaped troughs or depressions among the isobars, evidencing a whirl or cyclone with its axis horizontal. Now I would suggest that a horizontal cyclone is very like a cylinder electrical machine, with the earth acting as rubber and the upper regions of air acting as prime conductor, the air which has been charged by friction being discharged as soon as it is carried up to these higher regions and thus electrifying them continually until they locally discharge."

The effect of the electrification of the particles of water vapour is to cause them to unite into large drops. This phenomenon is often noticed in thundershowers in a heavy fall of large drops. An experiment due to Lord Rayleigh illustrates this in a beautiful manner. A little vertical fountain is produced by connecting a piece of rubber tubing provided with a glass nozzle to an ordinary water pipe or other suitable supply. When a piece of electrified sealing wax is brought near the fountain the finer drops coalesce into larger ones and the jet changes in form. Electrification of fine water vapour can thus cause clouds to deposit their vapour in the form of rain by bringing the fine particles together and by thus increasing the weight of the drops.

The northern light I have said can be supposed to be brushlike discharges in the higher rarefied regions to lower regions. The brush discharge is plainly seen about an ordinary electrical machine when we approach

---

* Lightning Conductors and Lightning Guards, Lodge, p. 3.

a finger to the conductors in a dark room. When we bring the finger near enough to the conductor we obtain a spark, and it is this phenomenon, and not the brush discharges, that we perceive in lower latitudes in the case of an ordinary thunderstorm.

The work done in these so-called silent discharges, like the aurora or in the ordinary brush discharge, is small compared with that done by the lightning discharge, as we can see roughly by the differences in the intensity of the light produced. I have tested this question of difference of work in the following manner: A Leyden jar, with its outer coating slit so as to produce alternate spaces of tin foil and glass, was charged to a sufficient degree to produce a spark between terminals a fixed distance apart. The spark was examined by a revolving mirror, and the number of oscillations or surgings to and fro was noted. At each discharge between the spark terminals a brush discharge occurred between the slits in the coating of the jar. When the jar was placed in oil this brush discharge ceased, but no essential diminution could be perceived in the energy manifested in the spark. The number of oscillations were the same, and the duration of the spark was not apparently modified.

CHAPTER XVII.

WAVE MOTION.

Our study of electricity leads us now to the general subject of wave motion, which up to the time of the laying of the Atlantic cable seemed to be very little in touch with practical life. It was a subject for mathematicians and the natural philosophers, and it seemed to have no commercial importance. In signalling, however, through the cable the practical man was speedily confronted with problems of wave motion, and with the invention of the telephone the study of wave motion became instantly of importance to the practical electrician. The progress of electricity is steadily in the direction of the economical production of wave motion.

"By a wave is understood a state of disturbances which is propagated from one part of a medium to another." Energy pauses, and not matter. Waves are free or forced. An example of a free wave is afforded by that of the wave running into the Bay of Fundy, which is almost free from the influence of the sun or moon; while the ocean tide is a forced wave, since it depends upon the continued action of the moon and sun.

It has been computed that waves on the ocean of about three hundred feet long travel at the rate of nearly forty feet per second, or twenty-seven miles per hour. Their disturbance, however, is merely superficial. Even if they are forty feet high, the disturbance of a water particle at a depth of three hundred feet is not quite half an inch from its mean position. The depths of the ocean are practically undisturbed by such waves on the surface (Prof. Tait).

Although the study of wave motions of heavy fluids, like water, or even air, may provide us with analogies by means of which we can illustrate wave motions in an attenuated medium like the ether, we must bear constantly in mind the fact that the viscosity of water or that of the air greatly modifies the circumstances of wave motion.

Our ideas, however, of waves in the ether of space, which are believed to convey the energy of the sun to us, are primarily obtained from contemplation of the wave motions which we perceive in water and the air. The electric spark has been used in an interesting manner to make manifest waves in air which otherwise would escape our senses. Prof. Boys * by its aid has photographed the waves caused by the motion of a bullet. His method is substantially as follows:

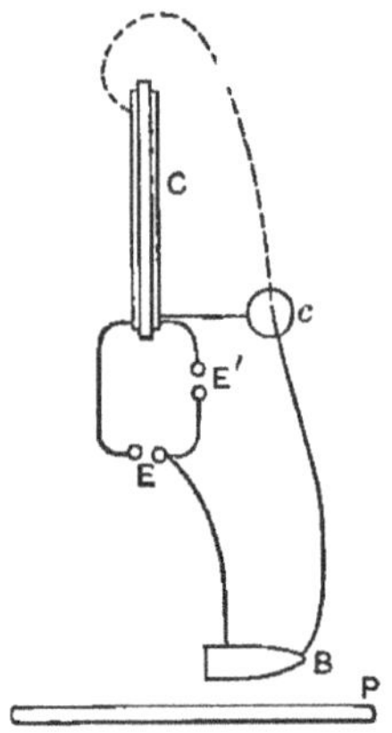

FIG. 30.

C is a plate of window glass (Fig. 30) with a square foot of tin foil on both sides. This constitutes the condenser, and it is charged until its potential is not sufficient to make a spark at each of the gaps, E and E′,

* Nature, March 9, 1893.

though it would, if either one of these were made to conduct, immediately cause a spark at the other; *c* is a Leyden jar of very small capacity connected with C by a wire—as shown by the continuous lines—and by a string wetted with a solution of chloride of calcium, as shown by the dotted line. So long as the gap at B is open this little condenser, which is kept at the same potential as the large condenser by means of the wire and wet string, is similarly unable to make sparks both at B and E′, but it could, if B was closed, at once discharge at E′. Now, suppose the bullet to join the wires at B, a minute spark is made at B and at E′ by the discharge of *c*. Immediately C, finding one of its gaps, E′, in a conducting state, discharges at E, making a brilliant spark which casts a shadow of the bullet upon the photographic plate, P. The wet string suffices to charge the jar *c*, but acts like an insulator when the discharge takes place at E′ and B. The photograph is a silhouette, but it serves to define the wave of air caused by the bullet.

Prof. Boys remarks that the wave revealed by the photograph shows just as in the case of waves produced by the motion of a ship, which become enormously more energetic as the velocity increases, and which at high velocities produce an effect of resistance to the motion of the ship far greater than that of skin friction, that the resistance which the bullet meets increases very rapidly when the velocity is raised beyond the point at which these waves begin to be formed. Scott Russell has shown by diagrams and experiments what happens when a solitary wave meets a vertical wall. As long as the wave makes an angle with the wall it is reflected perfectly, making an angle of incidence equal to the angle of reflection, and the reflected and incident waves are alike in all its parts. When the wave front nearly perpendicular to the wall runs along nearly parallel to it, it then ceases to be reflected at all. The part of the wave near the wall gathers strength; it gets higher, travels faster, and so causes the wave near the wall to run ahead of its proper position, producing a bend in the wave front, and this goes on until the wave near the wall becomes a breaker. To see if a similar phenomenon could be traced in the air, Prof. Boys arranged three reflecting surfaces (as seen in Fig. 31). Below the bullet two waves strike a reflector at a low angle, and they are perfectly reflected.

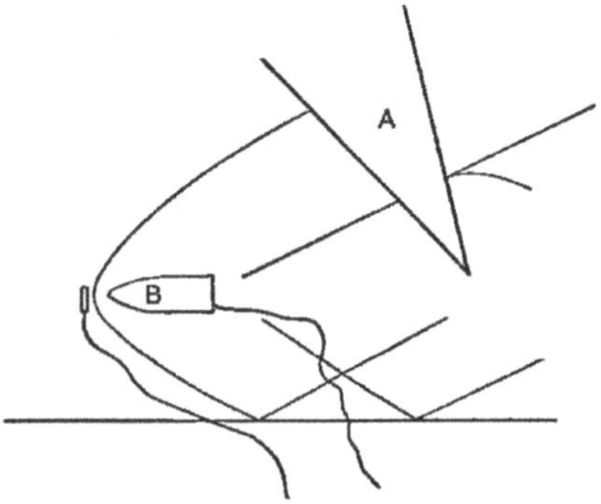

FIG. 31.

The left side of the V-shaped reflector was met at nearly grazing incidence. There is no reflection, but the wave near this reflector is of greater intensity; it has bent itself ahead of its proper position, just as the water wave was found to do. The stern wave has a piece cut out of it and bent up at the same angle. Prof. Boys points out that if the wave was a mere advancing thing the end of the bent-up piece would leave off suddenly, and the break in the direct wave would do the same. But according to Huyghens's hypothesis, the wave at any epoch is the resultant of all the disturbances which have started from all points of the wave front at any preceding epoch. The reflector, where it has cut this wave, may be considered as a series of points of disturbance arranged continuously on a line, each coming into operation just after the neighbour on one side and just before the neighbour on the other. The reflected wave is the envelope of a series of spheres beginning with a point at the place where the wave and the reflector cut, growing up to a finite sphere about the end of the reflector to a centre; beyond this there are no more centres of disturbance, the envelope of all the spheres projected upon the plate—that is, the photograph of the reflected wave—is not therefore a straight line, leaving off abruptly, but it curls round, dying gradually to nothing. In the nonreflection of the air wave by the V-shaped reflector we have optical evidence of what goes on in a whispering gallery. The sound is probably not reflected at all, but runs round almost on the surface of the wall from one part to another.

A most interesting method of studying sound waves in air by means of the electric spark was devised by Töpler.* He succeeded in making visible the reflection and the refraction of sound waves, and also the interference of two sound waves.

An idea of Töpler's method of rendering visible to the eye the waves of sound in the air can be obtained from a consideration of the phenomenon of mirage. A low-lying strata of air of suitable density enables us to see objects below the horizon, for the rays of light (Fig. 32) from these objects are bent or refracted to the eye by the strata of air. For instance, if A represents the position of the horizon, and S that of the sun, which is a little below the horizon, the strata of air lying above A can refract the ray S C to E, and looking along C E we shall see the sun apparently elevated

---

* Annalen der Physik und Chemie, 131, 1867, p. 180.

above the horizon to D. If, now, a sufficiently powerful sound wave could be generated on the horizon near A, the alternate condensations and rarefactions produced by the wave as it progressed upward in the atmosphere would have the effect of producing images of the sun at points D, D′, D_2, etc., and the eye at E would see in these images the progress of the wave through the air.

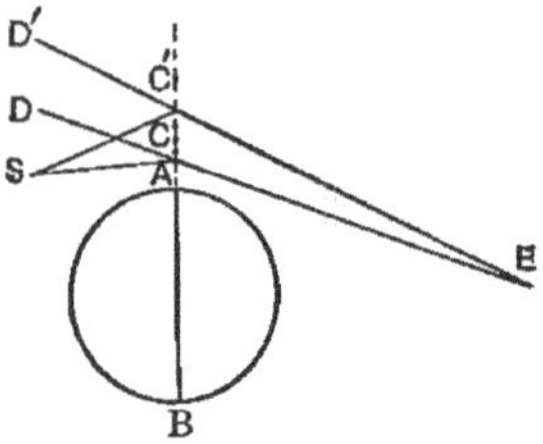

Fig. 32.

Of course we could not produce a sufficiently powerful sound wave to produce a mirage effect, and to thus imprint, so to speak, the condensation effects of the sound wave on the sky. The experiment, however, can be performed in the laboratory with an electric spark in place of the sun, and with a diaphragm, A B, instead of the earth. A certain arrangement of lenses also serves to make the phenomenon more definite.

The ease with which air transmits even a whisper is wonderful. The waves of sound pass through doors, are reflected from walls, and spread around innumerable obstacles, and still reach our ears. One is astonished at the readiness with which sound waves are reflected from tree-covered hillsides, and at the accuracy of the echo we hear. If the eye could follow a sound wave in the air as it passed in front of one, one would see spaces of compressed air and spaces of rarefied air. Although we can not see such spaces, we can represent the propagation of a sound wave so that its motion at any instant can be studied. Let the intervals of time be laid off along M X (Fig. 15, page 143), and the amount of compression or rarefaction of the atmosphere at each interval of time be laid off perpendicular to this line.

The varying compression will be expressed by the curve above the line M X, and the rarefaction by the portion below that line. The curve M *a b c d e* will then represent a sound wave. By an indirect process we can perceive the sound waves in air. A small air chamber, (Fig. 33), is provided on one side with a thin membrane, M; a little orifice, O, fits into this air chamber, and gas is led by the pipe to this chamber. With a suitable gas pressure and a suitable flame at O one can study the to-and-fro motions of the air at M by means of the motions of the gas flame seen in a revolving mirror, N.

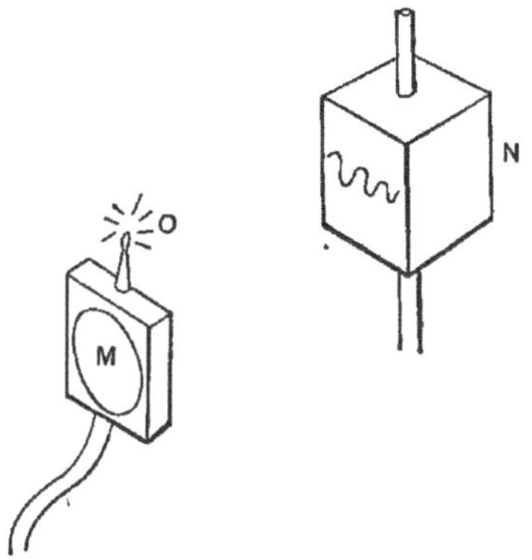

FIG. 33.

A modification of the experiment of using parabolic mirrors for the transmission of sound waves is interesting to show how extremely short sound waves may be detected. At the focus of the receiving mirror is placed a small box, the cover of which is replaced by a layer of thin bladder. Gas under suitable pressure is led by a small tube into this box, and a very fine orifice opposite the leading-in pipe is provided. On lighting the gas at this orifice we obtain a little flame which is extremely sensitive to high notes. On placing the box at the focus of the receiving mirror and shaking a bundle of keys or striking two pieces of metal together at the focus of the sending mirror the flame will lengthen and shorten. Experiments like this help us to realize how the air is filled with invisible waves of sound when we speak, and how they are reflected, spread behind obstacles, and brought to a focus by mirrors. We must, however, constantly bear in mind the fact that sound waves move to and fro along the direction in which they are propagated. It is thus only that ordinary air can transmit vibrations. It is not capable of transmitting transverse vibrations, or, in other words, vibrations at right angles, to the direction of propagation. It can not transmit, therefore, light and heat and electrical vibrations; they apparently require an ethereal medium for this.

We have seen that sound waves can be represented by sinuous curves which resemble waves in water or waves in the ether. We have said that the atmosphere can not transmit transverse vibrations, and that the to-and-fro movements in the sound wave are in the direction of its propagation. Now we should expect that the velocity of this to-and-fro movement would be greater in metals and liquids than in the air, since in solids and liquids the particles are nearer together and more numerous than in the atmosphere. A push would therefore be transmitted faster in the denser media. This is found to be the case. The velocity of sound in iron is ten or twelve times and in water four or five times that in air. The light and heat rays, however, travel slower in water than in air. The ether movements are impeded by the particles of gross matter.

The greater ease with which sound travels in water than in air has led to many attempts to signal great distances under water. No great measure of success has yet been obtained in this direction. By means of a microphonic attachment to a vibrating diaphragm I succeeded in hearing the clicking of two stones under water a distance of six hundred feet.

The simplest idea of wave motion which is characteristic of light and heat waves can be obtained from the waves which are propagated in water when a stone is thrown into a pond. The ripples extend in constantly widening circles from the centre of disturbance. They pass under bits of wood or cork, but do not urge these onward. They give an up-and-down motion to these bits of matter. In other words, the up-and-down motion of the waves is transverse to the direction of propagation of these waves. In this respect the waves in water resemble light and heat waves; for in these waves also the vibration is at right angles to the direction of propagation. If now we should be in a row-boat between the wakes of two steamboats, it is evident that if the wave from one steamboat should tend to urge our boat upward while the wave from the other should tend to depress it, we should escape being swamped, for the two waves would neutralize each other. On the contrary, if they should both combine to lift and to depress our boat, the danger would be greatly increased. There would be points of interference of waves and augmentation of waves.

In the subject of light and heat we are dealing with minute waves; and in order to observe the interferences of such minute waves we must use very small slits or orifices. If, for instance, we should look at the light of a candle through a slit in a card one tenth of an inch wide, we can scarcely perceive any phenomenon of interference. On the other hand, if we should cut a mere line in a thin piece of metal and look at the candle, we should perceive bright and dark lines extending each side of the narrow slit. These bright and dark spaces are due to the waves of light which emanate from the illuminated edges of the slit, and which interfere. In looking through an umbrella at a distant electric light we can also perceive phenomena of the interference of light in passing through the meshes of the covering of the umbrella.

Now, these waves in water and waves of light and heat in the ether differ essentially from the waves of sound. The atmosphere can not transmit transverse vibrations, such as we perceive in the ether. It can only transmit vibrations which are in the line of the onward movement of the wave. In the movement of a sound wave the air is alternately compressed and rarefied. It is beating like the human heart. Now in this peculiarity of air—that it can not transmit transverse vibrations—we have one of the strongest arguments for the existence of the ether medium; for we can prove that the vibrations producing light and heat are transverse to the direction of onward movement of the waves of light and heat.

The difficulties which waves of light and heat find in passing through liquid and crystalline substances can be explained by the effect of molecular arrangements in modifying the waves in the ether surrounding or interpenetrating such molecular groupings. For instance, a plate of the crystalline tourmaline will allow only such waves in the ether to pass through it whose transverse vibrations are confined to one plane. Its molecular grouping cuts off certain transverse vibrations. This grouping behaves to the waves, as we have already said (page 80), like the slats of a Venetian blind to a shower of stones which is thrown against it. Only the stones which are in the plane of the openings in the blind can pass through. The light waves are then said to be polarized; their vibrations are confined to one plane. If a ray of light which has thus passed through one plate of tourmaline should be examined by another plate of tourmaline, it will be found no light can be seen if the second plate is turned into a certain position. A second Venetian blind has been turned so that its slats are at right angles to the first, and the stones which passed through the first blind are intercepted by the second. Thus, two plates of a crystal which are separately transparent can completely stop light from passing through them if combined in the manner suggested. Sir John Herschel suggested that by means of this phenomenon one could telegraph through the air secretly. A beam of light could be transmitted through a plate of some substance like tourmaline, and be thus polarized. The distant observer, provided with tourmaline spectacles, could thus read the signals which would escape the observation of every one else. There are many substances besides tourmaline which possess the property of polarizing light, or, in other words, confining its wave motions in one direction or in one plane. When we study the subject of the polarization of light, we are led to a consideration of the effect of molecular groupings upon the transmission of the waves of light or heat.

In this chapter, however, let us confine ourselves to the consideration of wave motion, and avoid, for the present, a consideration of the entangling effect of atoms and molecules. Whether we accept the doctrine of the ether or not, we certainly see in the phenomena of light and heat evidence of a to-and-fro movement which is periodic—that is, it is like the motion of a point on the rim of a bicycle wheel: it rises to its highest point above the hub, and sinks to its lowest point below the hub, and goes through all the intermediate positions with every revolution of the wheel. Its greatest amplitude of movement above and below the horizontal line passing through the hub is the radius of the wheel. It goes through all its points in a certain period of time. To the eye standing behind the revolving wheel the motion of the point is like that of a boat which rises and falls with a wave. If the height of a boat should be plotted at progressive intervals of time while the wave passes under the boat, we should again obtain a periodic curve—periodic because it recurs in the same form. The length of the wave includes a crest and a trough (Fig. 34). If now our boat should be in the neighbourhood of a wall of rock, at a point, M, where the incoming wave and the wave reflected from the rock meet (Fig. 34), it would evidently not rise or fall. This point is called a node. It is a place where there is no motion. The other points on the wave belong to what are called ventral segments. They exist in the ether when a light wave meets in its progress to a reflecting surface the backcoming reflected wave. Since the lengths of the light waves are so extremely small—about $\frac{1}{40000}$ of an inch—it is evident that it would be extremely difficult to detect these nodal points in the ether.* With sound waves, however,

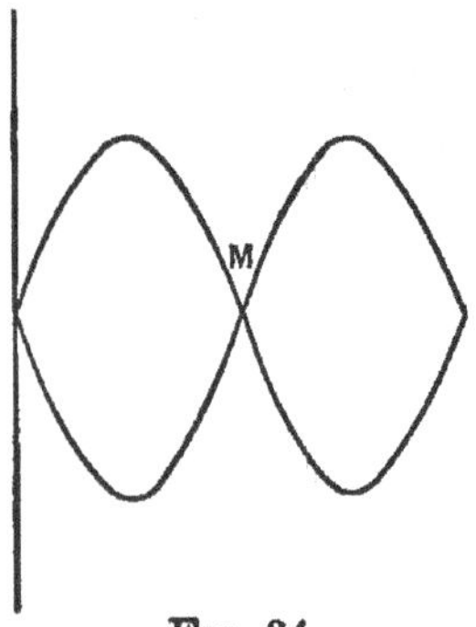

FIG. 34.

* They have, however, been photographed lately by Wiener (Annalen der Physik und Chemie, No. 40, 1890, p. 203), and a short consideration of his process will illustrate the formation of waves. Since the light waves are so extremely small, the thickness of a photographic film which could show the paths of the waves must be comparable with the wave length of the light. Wiener therefore used a film of collodion, sensitized by a chloride-of-silver solution. The thickness of the film was about $\frac{1}{20}$ to $\frac{1}{30}$ of the wave length of

which are several feet long, the nodal points can be readily made manifest. We shall see that apparently they can also be detected when electro-magnetic waves are reflected from the walls of a room.

Perhaps the most striking evidence that the phenomena of light and heat are due to wave motions is obtained from the phenomena of interference of waves; and since investigators are beginning to speak of an electrical spectrum, it is well to obtain an idea of the light and heat spectrum, which is really, according to our modern ideas, also an electrical spectrum.

If two pieces of perfectly plane glass are placed upon each other so as to include a thin film of air between their plane surfaces, beautiful bands of colour are seen when these surfaces are looked at obliquely. These colours are due to the interference of waves of light, and are produced in the same manner as the bands of colour on soap bubbles or the colours of thin films of oil on water. If a cent which has been slightly warmed in the fingers is placed on the upper surface of the upper glass plate, the coloured bands immediately shift their position and change from straight bands to curved ones. This effect is produced by the expansion of the glass

---

sodium light (about $\frac{1}{50000}$ of an inch). The light was reflected perpendicularly from a mirror. If the sensitive plate was placed perpendicular to the surface of the mirror, it is evident that the nodal points corresponding to M, Fig. 34, where the reflected wave crossed, so to speak, the incident wave, would lie in a series of planes parallel to the mirror. The distances, however, between such nodal points would be too small to observe on the photograph. If, now, the photographic plane is inclined at a small angle to the mirror, the apparent distances between the nodes is increased, and a series of dark bands can be obtained on the sensitive plate which show the nodal points and the neutral segments of the wave. In this way photographs of light waves are obtained.

due to the heat of the cent, and this expansion produces a variation in the thickness of the thin film of air between the glasses. This effect can be observed with glass plates half an inch in thickness.

The interference of the waves of light is also illustrated by Lippman's process of colour photography.

Lippman states that the essential conditions for obtaining colours in photography are two: First, continuity of the sensitive layer; second, presence of a reflecting surface against which the sensitive plate rests. By continuity of layer is meant the absence of grains. It is necessary that the iodide or the bromide of silver should be disseminated throughout the mass of a plate of albumen or gelatine in a uniform manner without forming grains which are of sensible size with reference to the dimensions of the waves of light. Lippman employs a support of albumen, collodion, or gelatine with the iodides and bromides of silver. The dry plate is carried in a tray in which one pours mercury. This mercury forms a reflecting surface in contact with the sensitive plate. The exposure, development, and fixing is the same as in the ordinary process, but when the plate is fixed and dried colours appear. The plate is negative by transmitted light and coloured by reflected light. The theory is as follows: The incident light which forms the image interferes with the light reflected from the surface of the mercury. There are therefore formed a system of interference bands—maxima and minima—in the interior of the sensitive layer. The maxima alone are impressed upon the layer, and remain marked by layers of silver more or less reflecting which take their place. The sensitive layer is thus divided by these layers into a series of thin plates, the thickness of which is equal to the interval which separates two maxima—that is to say, to a half wave length of the incident light. These thin layers have therefore precisely the necessary thickness to reproduce the incident light by reflection. The visible colours are thus of the nature of those one sees on soap bubbles. They are, however, purer and more brilliant if the photographic operations have given a good reflecting surface; for one forms in the thickness of the sensitive layer a very great number of thin plates superposed; about 200 of the layers amount to $\frac{1}{20}$ of a millimetre. The reflected colour is the purer the greater the number of reflecting surfaces (Comptes Rendus, 112, 1891). Lippman also applied the orthochromatic process to his method. With layers of albumen, sensitized by bromide of silver, rendered orthochromatic by azurine and cyanine, he obtained brilliant photographs of the solar spectrum. He exhibited to the French Academy a photograph of a stained window with four colours—red, green, blue, and yellow; a group of *drapeaux*, a plate of oranges surmounted by a paroquet. The *drapeaux* and the bird were exposed five to ten minutes under the electric light; the other objects were exposed several hours to different light. It is necessary, therefore, to greatly increase the sensitiveness of the plates.

Twenty years ago Nobert succeeded in ruling fine lines on glass which were separated by as small a distance as $\frac{1}{100000}$ of an inch. His lines were used as a test of the resolving power of microscopes, and his method of producing them was considered a secret. It was not long, however, before it was discovered that a diamond cutter could be moved by means of an accurately cut screw through very small spaces. Thus, if the distance between the threads of the screw is $\frac{1}{25}$ of an inch, and a circle forming one end of the screw were divided into one thousand parts, the movement of the head through one of its smallest divisions would move the diamond cutter connected with a nut on the screw through $\frac{1}{25000}$ of an inch. At one point a line could be cut and a parallel line $\frac{1}{25000}$ of an inch from this, and so on. Success in forming a good diffraction grating—for so this assemblages of fine lines was called—consisted merely in using an accurate screw and a plane surface.

If a source of light (Fig. 35) is placed at D, the rays reflected by the sides of the diamond cuts B and C (greatly magnified in the drawing) will pursue paths A C and A B, which may differ by an odd number of half wave lengths, in which case they will neutralize each other at A, and there will be no light, or else they may differ by an even number of wave lengths, in which case there will be light at A. We shall thus have a spectra interspersed with dark spaces along a line passing through A. Instead of using lines ruled by a diamond on a glass plate and transmitted light, we can employ such lines on speculum metal, and employ reflected light; in which case the light appears to come from a point behind the mirror, as shown in the lower part of Fig. 35. Another way to produce interference is due to Prof. Michelson. He has made two important uses of his apparatus.

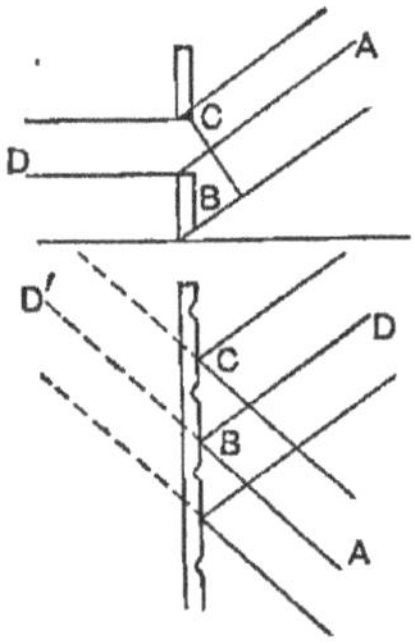

FIG. 35.

A source of light, S (Fig. 36), is reflected by a plane parallel glass, B, to a plane mirror at A. The latter in turn refracts it through the glass to D. Another beam is reflected to C, and then back to D. There is therefore a difference in the paths of the rays reflected from A and C, and interference takes place at D in the shape of a great number of fine dark lines. Prof. Michelson used a modification of this apparatus to test the question whether the ether moves with the earth. The apparatus was set up in such a way that the path of the light A C was in the direction of the movement of the earth. The mirror A therefore is carried to A′, while the reflected light travels to B and returns. If the ether did not move with the earth the position of interference bands would indicate this. The apparatus was set up with the greatest care at Potsdam, near Berlin, but no effect could be noticed. Prof. Michelson has lately, by means of this method, based the measurement of a standard of length on the measurement of the wave length of light. If the paths B A or A C are altered, this change can be measured by the number of interference bands which cross the field of view at D. The change in length can thus be estimated in a definite number of wave lengths of light. The tenth of a metre (or the decimetre) can thus be expressed in wave lengths. The advantage of this method of measurement consists in using an unchangeable standard instead of the length of King Henry's arm. The imperial yard or metre is subject to expansion and contraction, and doubtless to secular change. Prof. Michelson's standard is apparently immutable.

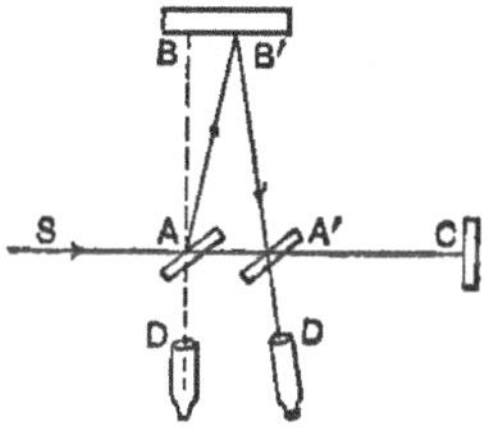

FIG. 36.

The Rowland concave grating is the greatest contribution that has been made to the subject of spectrum analysis since Rutherford showed the possibility of making diffraction gratings which would give spectra of great brilliancy and large dispersion. Instead of using a plane surface ruled with fine lines, Rowland employs the surface of a concave mirror of long focus. The waves of light in striking these fine lines are reflected and brought to a focus by the concave mirror.

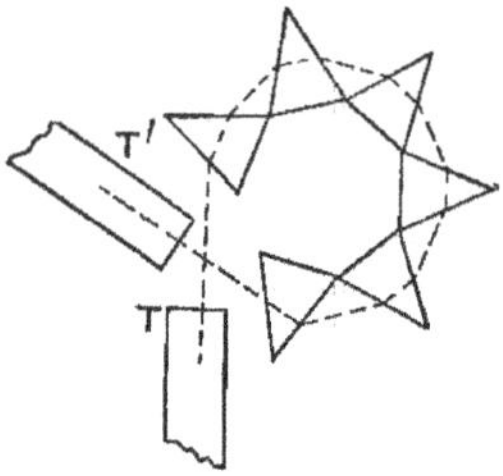

Fig. 37.

The modern spectroscope forms a striking contrast to the large spectroscopes which were used twenty years ago. In the latter form the light was sent through a great number of prisms to refract and disperse it to the utmost. In some cases the light after being refracted by one set of prisms (Fig. 37) was sent up one story by means of plane mirrors, and was then sent round another set of prisms. It is evident that the materials of the prisms modified to a great extent the amount and character of the dispersion obtained. The modern spectroscope (Fig. 38) consists simply of an illuminated slit, S, a concave mirror, C, ruled with fine lines, and an eyepiece or photographic plate at E.

It is interesting to reflect that the solar spectrum was before the eyes of men like Sir Isaac Newton and Goethe, with its intimations of the inner mysteries of the sun and stars, yet these mysteries were veiled from such men. For many years after Bunsen and Kirchhoff pointed out the road to spectrum analysis the spectrum was the subject of human inquiry, yet it contained other mysteries the clews to which have been slowly grasped even by the best minds. The fact that the distinction between light rays and heat rays resides only in a difference of wave length has slowly dawned upon human

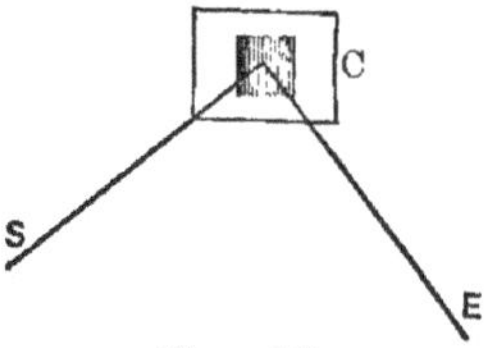

Fig. 38.

intelligence. It was not suspected twenty years ago that the spectrum of the invisible rays which we call heat rays extends to a length exceeding the visible spectrum ten or twelve times.

To the labours of Prof. Langley we are indebted to a great extension of our knowledge of the invisible heat rays. The method he employed to detect these rays is an interesting example of the transformation of energy. The instrument by means of which he opened up this undiscovered country he terms a bolometer—from βολη, throw, and μετρον, a measure. It consists of a very fine wire through which circulates a steady current of electricity. A very slight change in temperature of this fine wire will cause a change in its resistance, and a quick movement of a galvanometer needle connected with this wire will result. This throw of the galvanometer needle is taken as a measure of the amount of heat which falls on the fine wire when the spectrum is moved across such a wire. The galvanometer connected with the wire will show a cooling effect when a dark solar line passes across the wire. In this way Prof. Langley has detected and mapped thousands of lines in the invisible red, where only a dozen were known to exist before his labours.

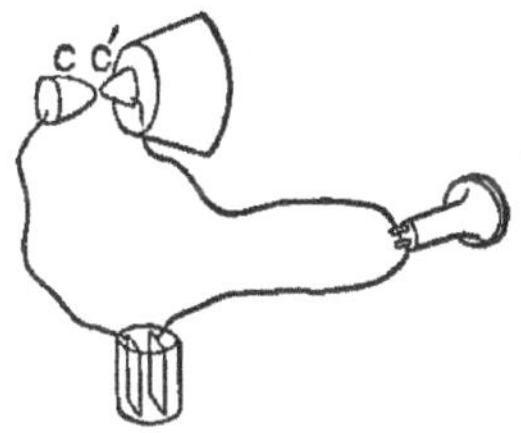

FIG. 39.

I have said that the bolometer is an interesting example of the transformation of energy. While it is being heated by the electric current passing through it, it is so sensitive to the temperature of outside bodies that it serves as a measure of the energy it receives from these bodies. A heat wave may be said to be transformed into an electric current to register itself.

By the aid of the bolometer Langley has been enabled to measure the heat of the moon and to examine its spectrum. The actual arrangement of the bolometer apparatus is that of the Wheatstones bridge (Fig. 40).

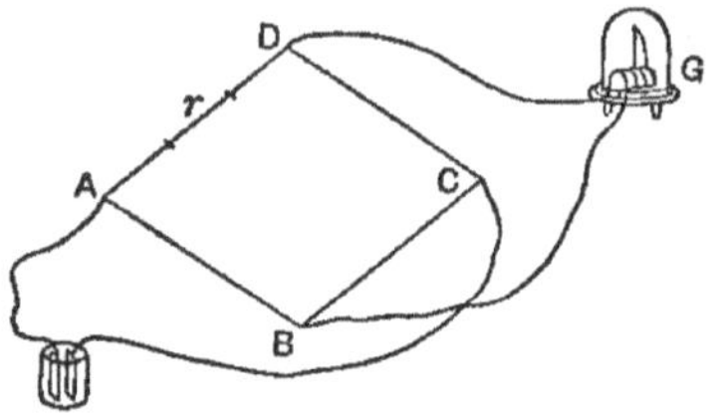

FIG. 40.

The action of a fine wire which is traversed by an electric current toward radiant heat is very analogous to that of the carbon transmitter toward sound waves. If we connect two carbon points (Fig. 39) in circuit with a battery and affix further one of these points to a vibrating diaphragm, we have practically the modern carbon transmitter. The voice changes the resistance at the carbon points and makes the electric current fluctuate in unison in the distant telephone. The resistance of the bolometer also changes with the waves of radiant heat which fall upon it. In both the carbon transmitter and the bolometer the energy of wave motion is changed into electrical manifestations.

Can not we therefore speak to a fine wire and use its fluctuations of resistance to measure the energy with which we speak? This is possible. We have only to supplant the galvanometer, G, in Fig. 40, by a telephone. The fluctuating currents produced in the telephone by the human voice disturb the distribution of the electrical currents through the sides of the parallelogram, of which the bolometer wire, A D, makes one portion, but how shall we obtain evidence of this transformation? One of the best methods is due to Profs. Rubens and Arons, who have modified the distribution of currents in the Wheatstones bridge. Parallelograms of fine iron wire, W_1 and W_2 (Fig. 41), take the place of the usual resistances between A D and D C (Fig. 40). If one connects a telephone to the points B and D (Fig. 41) and speaks into it, the distribution of the currents along the sides of the little parallelogram W_2 is disturbed. This disturbance leads to a disturbance between D and B (Fig. 40), which is made evident by the movement of the galvanometer needle G. The voice thus alters the circulation of electricity through this network of conductors.

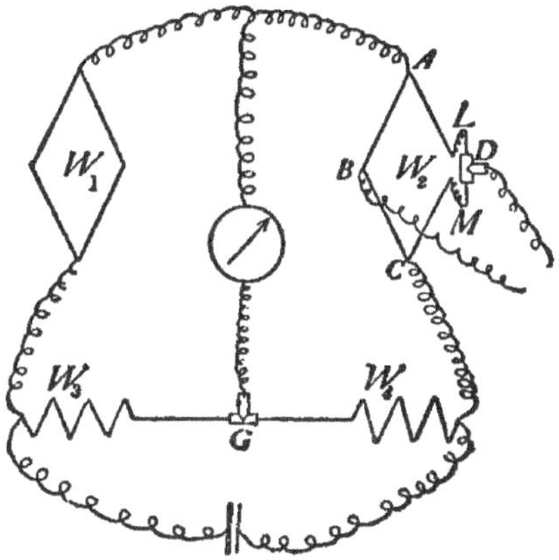

FIG. 41.

We have now used the fluctuations of electrical resistance to make manifest the waves of heat and light and sound. Can we also use this instrument for the detection of electrical waves? This has been accomplished by Prof. Rubens and Prof. Arons, who supplanted the telephone in Fig. 41 by loops of wire, O and P, on glass tubes (Fig. 42) which could be slipped along the wires traversed by the electrical waves. The fluctuations of electricity at electrical ventral segments and nodes thus caused corresponding

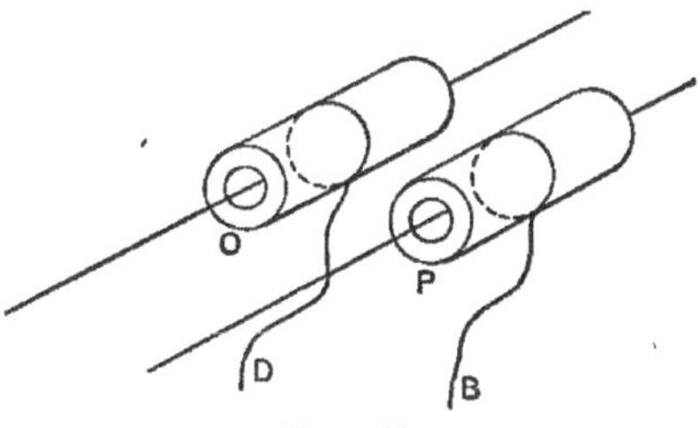

FIG. 42.

disturbances through the glass tubes on the terminals O and P, which in turn disturbed the distribution of electrical currents in the parallelogram W_2. The little glass tubes are really Leyden jars. The portion of the wire in the tubes along which the electrical waves are propagated constitute the inner coating of these jars, and the loops O and P the outer coatings.

In a previous chapter on the rates of change of magnetic induction; we have shown how varied are the transformations of energy which can be effected by the quivering, so to speak, of lines of magnetic flow. In the use of these little Leyden jars we perceive that the electrostatic lines can also, by their rate of change, effect a rate of change of magnetic lines of flow. The fluctuation of the electrical charge on the coatings of the little jars produces a current on the wire connecting them. Indeed, if this fluctuation is accompanied by a sufficiently powerful electro-motive force, we have seen that it can produce the usual evidence of an electric current—a spark. On the other hand, if we should speak into the telephone in Prof. Rubens's apparatus, we ought theoretically to obtain an electrostatic disturbance on the little Leyden jars—that is, the fluctuations of the magnetic induction produced in the telephone by the human voice should be transformed into fluctuations of electrostatic lines between the coatings of the Leyden jars.

The effect would be small, but it would be possible. This effect, indeed, has been shown substantially by Dolbear, whose telephone is a Leyden jar to which one listens while a fluctuation on the wires connecting its two coatings is caused by the human voice acting on a carbon transmitter placed in the primary circuit of a Ruhmkorff coil, while the Leyden jar, of which the dielectric is air, is connected with the terminals of the secondary coil of the Ruhmkorff.

Our conceptions of the energy and rapidity of the changes which can be produced by fluctuations of the flow of magnetic induction and of electrostatic induction have been greatly enhanced by the invention of the telephone. And we are now, even in practical applications of electricity, obtaining a realizing sense of the importance of arranging our electrical circuits so that the waves which produce the fluctuations of magnetic and electrostatic induction should have their proper expression in the transformation we desire to accomplish. The scientific imagination even looks forward to transmitting intelligence from America to Japan by suitably modifying the electric charge on the earth. This does not seem at first sight more improbable than the feat of speaking by means of the waves of heat, which is accomplished by Graham Bell's photophone. The following interesting modification of this instrument has lately come to my attention:

A tube of lampblack, T (Fig. 43), is placed at the focus of a large parabolic mirror, and a speaking tube, or rather a listening tube, is connected with it.

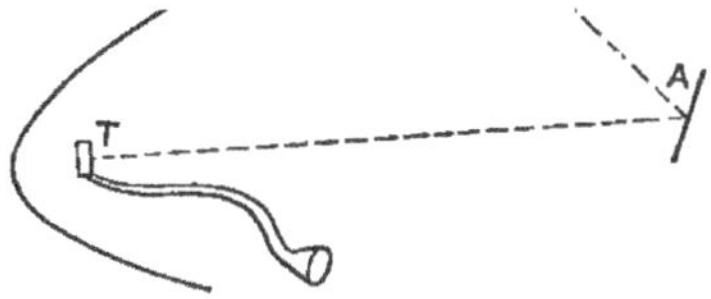

Fig. 43.

At the distance of half a mile a reflecting mirror is supported so that the rays of the sun shall be reflected to the tube, T. The voice of a speaker at A impinging against the mirror can be thus heard, and conversation carried on through the air. The heat rays are thrown into vibration by means of the speaker's voice upon the lampblack, and the air in the speaking tube is thus tuned to his voice. Here we have a transformation of sound movements into heat movements and a retransformation into sound waves. Can we not transform the sound waves of the voice into light waves, so that a speaker can cause light to appear at a great distance from him? To accomplish this, we need only to set the electrostatic lines of force to quivering in unison with the voice.

CHAPTER XVIII.

ELECTRIC WAVES.

When we survey the practical development of electricity, which I have outlined, we are struck with the fact that our minds have been led from a consideration of steady currents of electricity and the phenomena produced by them to what may be termed unsteady or periodic currents. The transformations of energy which are possible with periodic or alternating currents are far more varied than those we can accomplish with steady currents. It would seem that even the development of the applications of the alternating current suggests the electro-magnetic theory of light. The swifter the rate of alternation of our alternating dynamo, the nearer we approach to the manifestations of light and the more varied become the electrical phenomena. This seems to me the most remarkable conclusion to be drawn from Tesla's experiments on high frequency discharges. If we could excite electrical currents which would oscillate some billions of times a second we might produce the sensation of light on the retina of the eye without a spark.

The ordinary Leyden jar is the swiftest alternating machine which we can use at present. Joseph Henry showed conclusively, in 1840, that the discharge of a condenser is, in general, oscillatory. His observations on this oscillation form an epoch in the study of electricity, and the attention of the scientific world is now closely directed to the manifestations which he discovered, and which Lord Kelvin in 1850 expressed in a mathematical law which forms the basis of Hertz's celebrated work.

The rate of alternation of the Leyden-jar machine depends upon the capacity of the jar and the self-induction of the wire connecting its outer coating with the inner coating. The swiftest rate of alternation we can obtain from an alternating dynamo is barely one hundred thousand alternations per second, and this rate has practically not been reached. With a Leyden-jar discharge we can obtain and make evident by photography the rate of ten millions per second. The lightning discharge is a discharge from the Leyden jar formed by the layers of cloud and the earth, and its destructive effect is in part due to its rapidity of oscillation.

Now two Leyden jars with their circuits of wire can be electrically tuned so as to be in unison with each other, and when one jar is discharged the neighbouring one, which has not been charged, will also give a spark arising from what is termed electrical resonance.

It is important that we should obtain a clear idea of what may be termed electrical tuning or the obtaining of resonance. Let us, in the first place, examine what is termed resonance in the subject of sound. One of the simplest methods of showing acoustical resonance is to sound a tuning fork over the mouth of a long vertical jar and then slowly pour water into the jar. At certain points the column of air in the tube will vibrate in resonance with the tuning fork and a great augmentation of sound results. The particles of air swing in tune with the prongs of the fork. The best way of showing this phenomenon is to immerse a large glass tube open at both ends in a larger glass jar which is filled with water. By moving the inner glass tube up or down one can lengthen or shorten the column of air at pleasure and thus tune it to the fork. Another simple method showing the effect of resonance consists in mounting two forks which give the same note on hollow boxes closed at one end and open at the other, and placing these boxes with their open ends close together. If the two forks are exactly in tune when one is excited by a violin bow the other will respond, and will continue to vibrate when the exciting fork is brought to rest. If the forks are not in tune they can be brought into resonance by loading the prongs of one of the forks with a little piece of wax. Notes suitably struck on any stringed musical instrument can excite similar notes on another stringed instrument. Electrical tuning or the obtaining of electrical resonance depends upon conditions analogous to those which obtain in the subject of acoustics, and can be illustrated best by considering the photographs of electric sparks (Plate I, frontispiece).

Let us now endeavour to arrange the electrical circuits which shall be in resonance. It will be necessary first to obtain the time of oscillation of one circuit, and then arrange another circuit which shall have the same time of oscillation; the latter circuit will then be in resonance with the first circuit. One obvious way to accomplish this would be to discharge a Leyden jar through the circuit and photograph the spark which is thus produced by means of a revolving concave mirror. Knowing the speed of the revolving mirror and the distance to the sensitive plate, we can obtain the time of vibration of the circuit, and then we can arrange another electrical circuit which will have the same number of vibrations. If these two circuits are then placed parallel to each other, even twenty feet apart, a spark through one circuit will excite a spark in the other circuit. A simple way of arranging this experiment is as follows: An electrical machine is employed to charge a Leyden jar, A (Fig. 44). The accumulated charge equalizes itself between *m* and *n* around the cir-

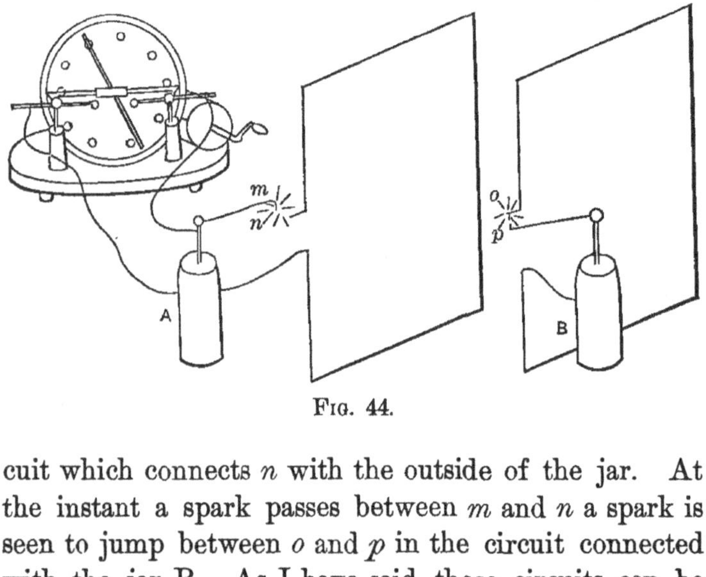

FIG. 44.

cuit which connects *n* with the outside of the jar. At the instant a spark passes between *m* and *n* a spark is seen to jump between *o* and *p* in the circuit connected with the jar B. As I have said, these circuits can be placed from ten to twenty feet apart, and can be made to respond to each other. Two principal factors enter into the phenomenon of electrical resonance: the arrangement of wire in the coils which are opposed to each other, and the number of Leyden jars—or, in other words, the amount of capacity which is connected with these coils. Thus in the above experiment we have the coils and the Leyden jars. The latter serve to accumulate the charge of electricity and to discharge it through the coils.

When we see an electric spark we must reflect that a magnetic wave reaches our eyes at the same instant as the light. Its velocity in the ether is the same as that of the light rays. When a spark occurs in one circuit a spark will also occur in another circuit; it may be across the room, if the latter circuit is parallel to the first circuit and properly tuned to this circuit. The energy of the first spark is conveyed through the ether in magnetic waves to the second circuit. The second spark appears apparently at the same instant as the exciting spark. The velocity of propagation of the magnetic waves which produce the spark is probably the velocity of light. The velocity of electricity should be measured in free space, and not on conductors, for on metals its propagation is retarded, and it takes time for the current to arrive at its greatest strength. Thus, if we have two parallel circuits with a battery and key in one of the circuits, and if we touch the key so as to send a current along one circuit, we cause lines of force to spread out in circles from the circuit, and these lines cause a current in the neighbouring wire in the opposite direction to the exciting current. The circles of force emanating from this second circuit embrace the first circuit and set up a current in it opposed to the exciting current.

In a paper on the oscillations of lightning discharges,* I expressed the opinion that the method first employed by Spottiswoode, of exciting a Ruhmkorff coil or transformer by means of an alternating-current dynamo, put in the hands of an experimenter a far

* Phil. Mag., October, 1893.

more powerful method of studying electrical oscillations than the old method of charging Leyden jars by means of an electric machine or by the use of a Ruhmkorff coil with a battery. I have therefore employed an alternating machine capable of giving 120 volts and a current of from 15 to 25 ampères, and have employed suitable transformers to obtain the necessary difference of potential to produce the sparks which I wished to study.

Generally I have employed one primary or exciting circuit between two entirely separate and disconnecting resonating or secondary circuits. The image of the three sparks thus produced could then be compared upon the same plate.

Without entering into a more detailed account of the apparatus I employed, I will state the most striking results which I have obtained. A unidirectional spark (nonoscillatory) always excites an oscillatory discharge in a secondary circuit if the self-induction, capacity, and resistance of this secondary circuit permit an oscillatory movement. It is therefore not necessary that the spark in a primary circuit should be an oscillating one in order to excite oscillations in a neighbouring conductor. In this respect two electrical circuits are not in close analogy with two tuning forks. It is difficult by a unidirectional movement of the prongs of one tuning fork to excite the vibrations of another fork which is not in tune with the first fork. In every secondary circuit, or circuits neighbouring to the primary circuit, the first effect of the exciting unidirectional primary spark is to make the secondary circuits act as if there were no capacity in their circuits. In these circuits a threadlike spark results which is exactly like that produced when all the capacity in the secondary circuits is removed. After a short interval of time the electricity rushes into the condensers and begins to oscillate, the strength of the oscillations rising, after one or two vibrations, to a maximum and then decreasing; the rate of oscillation finally assumes a steady state. The electricity seems to be separated only along the wires at first, and the circuit vibrates more like a closed organ pipe than an open one.

If a unidirectional primary spark excites oscillations in neighbouring circuits which are slightly out of tune, the phenomenon of electrical beats or interferences can be produced in these circuits, and can be shown by photography.

If the primary spark ceases to be unidirectional and is allowed to oscillate, the oscillations of the primary spark tend to compel those of the secondary or neighbouring circuits to follow them; if they are not sufficiently powerful to do this, they beat with the oscillation of the secondary circuit. Moreover, if all capacity is removed from the neighbouring circuits, they oscillate in tune with the primary circuit, following the latter exactly. The secondary circuits without capacity act like sensitive plates and exactly reproduce every disturbance in the primary oscillating circuit.

In Fig. B, frontispiece, S′ represents photographs of the unidirectional primary spark. S is the unidirectional spark produced in a neighbouring circuit, B, from which the capacity has been removed. S″ is the oscillating spark in the circuit C; the condenser of this circuit was an air condenser. The spark S shows that no oscillation is concealed by the heavy pilot spark of the exciting spark S′. The photographs S″ show that the unidirectional spark S′ can set the circuit C into oscillatory movement, and that this oscillatory movement continues long after the exciting blow has ceased. A careful study of many photographs of this nature shows that a circuit containing capacity and self-induction acts at the first instant as if no capacity were in the circuit. It then begins to oscillate with a higher period than it afterward reaches, acting at first like a closed organ pipe and subsequently like a pipe open at both ends.

In Fig. A, frontispiece, S′ represents again the oscillating primary circuit, S the oscillating secondary circuit C. The circuits are nearly in geometrical resonance. Slight beats, however, can be observed. The duration of the secondary is nearly the same as that of the primary.

As we advance in our study of the transformations of electricity we perceive that we are driven off, so to

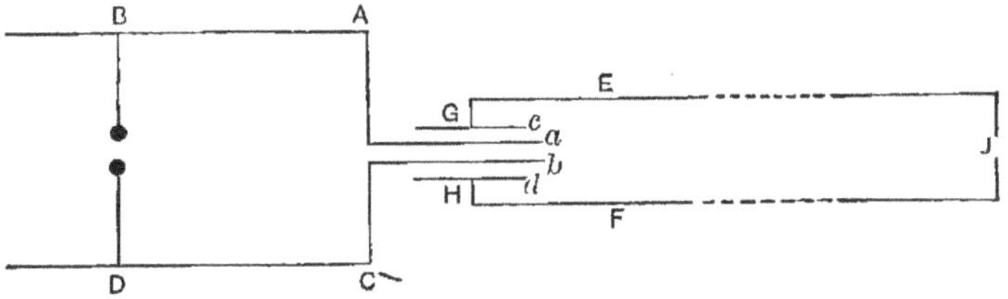

FIG. 45.

speak, from wires and conductors into the ether. The electrical manifestations refuse to show themselves on the conductors, except on the extreme outer layers of such conductors, while their most vigorous effects are displayed in the ether of space. We have, moreover, directed most of our attention to the effects at the terminals or ends of conductors. Let us now see if we can detect any form of wave motion along the wires or conductors. Returning to the use of a Ruhmkorff coil or step-up transformer, let us arrange two large plate condensers, *a* and *b*, parallel to each other (Fig. 45) and connect them with the terminals of the step-up transformer, providing a spark gap between B and D; then place two other smaller plates, *c* and *d*, opposite the plates *a* and *b*, and run long wires from these smaller plates fifty or sixty feet away to a spark gap at J. When a spark jumps across the gap between B and D a spark will also jump at J. This latter spark is due to the rate of change of the electrostatic lines between *a* and *b*. Now, on walking near the wires E F and applying the ear very close to them, but not touching them, one can find a point where a peculiar crackling sound is loudest; from this point the sound fades away in both directions. We have evidently detected a wave of electricity on the wires.

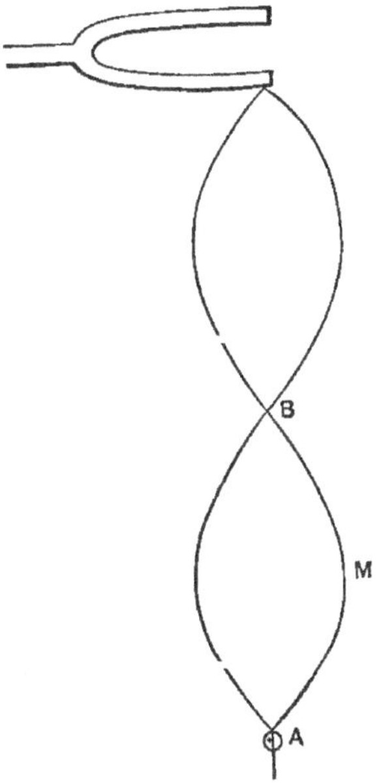

FIG. 46.

Let us now see if we can make it evident to the eyes instead of the ears. Taking a glass tube which has been rarefied to a great degree, let us rest it on the two parallel wires and move it along them. When it rests at the place where our ears detected the greatest sound, the tube lights up; the molecules in it are set into rapid movement. As we move the tube along the wires we find that the brilliancy of its lighting up diminishes as we go in either direction. We have evidently made manifest to the eyes an electrical wave. Let us return to acoustical analogies. If we should connect a silk thread to one prong of a tuning fork (Fig. 46), and support the other end on a cylinder at A, and suitably weight the string, it will vibrate with the fork. It is necessary that the time of vibration of the fork should be arranged with reference to the length of the string; in other words, the string must be tuned to the fork. Now, if we should touch the string at a node B we do not disturb the wave form on the string. If, however, we should touch even by a feather the vibrating portions of the string, we will say at M, the wave form can not be re-established; it is broken up. In the same manner, by suitably lengthening or shortening the wires E F (Fig. 45), we can find places where a conducting wire can bridge the two wires and not impair the brilliancy of the light in the exhausted tube. This conducting wire is then placed at the electrical nodes. In examining the conditions of obtaining electrical waves, we find that there are two principal conditions to be observed: the number of lines of force or magnetic ripples which emanate in rapidly expanding circles from every unit of length of the wires E F, and which are thus thrust into the space between the parallel wires E and F, and also the number of lines of electrostatic force between *a* and *b* and *c* and *d*, not to speak of the electrostatic lines which extend from every unit of length of each of the parallel wires. We find by further study that the time of vibration of the spark at J, or, in other words, the time of electrical surging along the wires, is proportioned to the square root of the product of the magnetic lines and the electrostatic lines per unit length. The spark at B D can be said to be the tuning fork which maintains the waves along the wires. The spark can be likened to a tuning fork in resonance with the fork at B D. If we could now photograph the spark at J by means of a rapidly revolving mirror, and thus spread out its vibrations so that they could be measured, we could obtain the time of oscillation of the electrical waves along the wires E F; if at the same time we measure the distance between the electrical nodes we should get half the wave length. Thus obtaining the wave length, we could obtain the velocity of propagation of the electricity along the wires. For the distance, l, which we call a wave length, is traversed with the velocity, v, of electricity in the time, t, or, to express this by an equation, $l = v\ t$. Thus measuring l and t we can obtain v. In a subsequent chapter I shall show how this has been accomplished, and how a number has been obtained for v which is very close to that of light—about 186,000 miles per second.

In the experiment we have described the wave motion of electricity has been apparently confined to wires or conductors. The question naturally arises, Can this wave motion be transmitted through space without wires? Hertz has shown how to detect these electric waves in the air by means of a circle of wire, the ends of which terminate in a micrometer screw (Fig 47), by means of which one can measure a very small spark gap. The dimensions of this circle are so chosen that its time of electrical vibration is the same as that of the circuit containing the exciting spark.

On moving along a straight horizontal line extending from the middle point, M, between the spark gap of the exciter (Fig. 47), a spark will appear at the spark gap, N, of the resonating circle. If now the resonating circle be moved to and fro between the spark gap M and a large metallic plane surface, S, certain nodal points can be found at which the spark in the resonator disappears. We have to do with electrical waves which emanate from the exciter, strike the metallic plane surface, and are reflected by the surface. The phenomenon is similar to that we should obtain if, having excited a powerful tuning fork some feet from a smooth

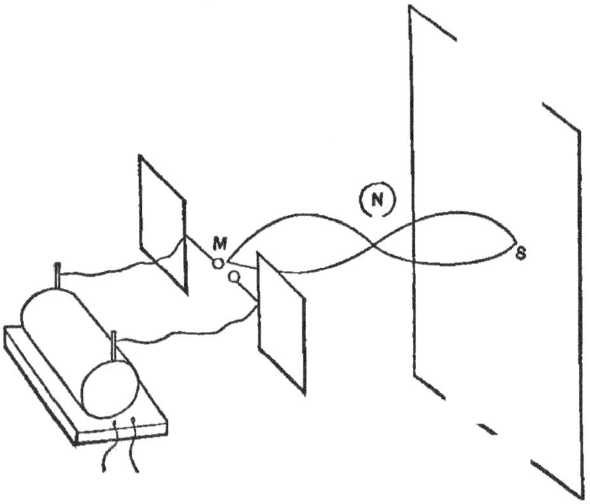

FIG. 47.

wall, we should obtain evidences of nodal points between the prongs of the fork and the wall. A simple way to do this is to walk toward the wall with the fork while it is sounding and note that there is a point where the sound of the fork becomes louder. This is where the reflected wave of sound re-enforces the movement of the prong of the fork. The column of air between the fork and the wall then vibrates in time with the prong of the fork, just as in the experiment when the fork is held over a cylinder which is moved up and down in water until a resonating column of air is obtained. Sarasin and De la Rive, working in a large room at Geneva, obtained reflected electric waves by the use of the circular resonator which we have described. The wave motion of electricity has thus been traced in the air, or rather in the ether, free from all conductors.

A still more striking way of showing that electrical waves can be reflected is due to Hertz, who used parabolic mirrors. Before speaking of the use of parabolic mirrors to transmit and receive electrical vibrations, let us examine their use in the subjects sound, heat, and light. If a watch is held at the focus of a parabolic mirror, A (Fig. 48), and a listener should station himself at the focus, B, of a similar mirror at a considerable distance, he can hear the tick of the watch. If a more powerful source of sound were placed at A, a sound inaudible at the distance of forty feet can be heard distinctly by the aid of the second mirror. The sound waves emanating from the focus A converge to the focus B of the receiving mirror. The light of a candle placed at A will also be reflected to the focus B; and the heat of a metallic ball placed at A can be

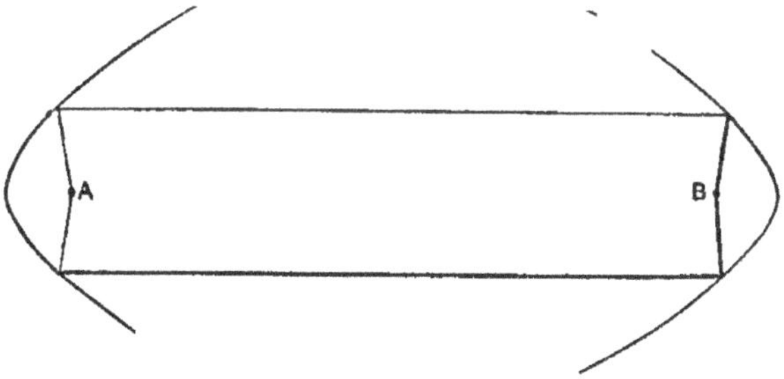

FIG. 48.

detected when the focus B is at least one hundred feet from A. It only remains to see if electrical vibrations emanating from the focus A can be detected also at B. Hertz has shown that this is possible. A simple method of constructing his mirrors is the following: A framework of wood is made, and parabolic curves, cut out of wood, are nailed to this framework. Over these parabolic guides stiff cardboard is nailed and is then covered with tin foil. In this manner one can make large mirrors of which every horizontal section is a parabola. With this species of parabolic mirror we have a linear focus instead of a focus at a point. If now an oscillatory spark is formed at the focus, A, of one mirror, electrical waves are sent out which are reflected from the surface of the mirror to the surface of the second parabolic mirror, and these converge to the linear focus B. If, then, a suitable conductor, including a spark gap, is placed along this linear focus, the waves decay along this conductor, and a spark is obtained as an evidence of this decay at the spark gap B. When B was placed forty feet from A, sparks could still be detected at B. It is probable that this distance can be greatly exceeded, and the imagination immediately pictures the possibility of sending and receiving electrical waves through a fog between one steamship and another. When the mirrors are so far apart that no evidence of a spark can be obtained at B, the focus of the receiving mirror, the waves can still be detected by an interesting process of transformation of energy. One terminal of a galvanometer of great resistance is connected to one side of the spark gap at B, and the other to its opposite side. When the electrical waves are received on the conductors at B a sufficient disturbance of the electrical state is caused to produce a slight electrical current through the galvanometer. Prof. Lodge employed another device, somewhat similar, to detect electrical waves. A glass tube is filled with metallic filings and is connected through a galvanometer with a battery. The inclosed air prevents electrical contacts. When electrical waves fall on the tube they cause minute sparks among the filings, and the battery is connected by these sparks with the galvanometer. He calls such tubes coherent tubes. The sparks serve the purpose of a relay to throw a stronger electrical impulse through the galvanometer.

Greater refinements can doubtless be made in apparatus to detect electrical waves. Indeed, our present form of apparatus will doubtless appear to the worker fifty years from now much as the rude mirrors of the ancients now appear to us. We have now given evidence that electrical waves can be reflected like light and heat waves; it remains to see if they can be refracted also. This refraction has been accomplished by Hertz, who placed a large prism of pitch between the parabolic mirrors, and found that, in order to obtain evidences of electrical waves at the focus of the receiving mirror, it had to be moved in order to receive the waves. We owe to Prof. Rhigi, of Bologna, a great simplification of Hertz's apparatus, and it is interesting to reflect that by means of this simplification we are brought back to the use of the electrical machine. We are again in the position of Benjamin Franklin in respect to apparatus. With great experience gained by means of the researches of Galvani, Volta, and Faraday, we have been led back by a distinguished Italian to the study of electricity in its most untrammelled manifestation. Instead of a Ruhmkorff coil or a transformer excited by a battery, Rhigi uses a small electrical machine such as we have shown (Fig. 2, page 27) in comparison with a Franklin machine. Two little spheres in oil (B and C, Fig. 49) receive their charges from the prime conductors, P and P′, of the electrical machine. Three sparks are formed, but the middle spark, B C, is the oscillatory one of very rapid period; for the capacity of B and C is very small, and so also is the self-induction of the little portion of the circuit, A B C D. The dimensions of the spheres and the circuit can be made so small * that electric waves of six millimetres (about

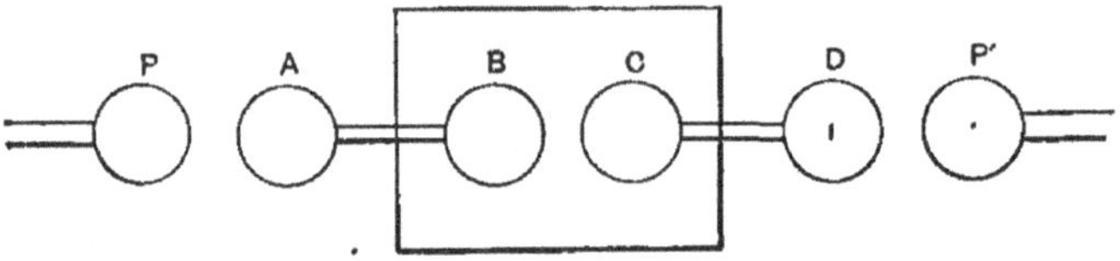

FIG. 49.

one quarter of an inch) can be sent out from this oscillator, whereas the shortest waves obtained by Hertz were several feet long. In order to detect these waves, it was necessary also to have a resonator or electrical eye of very small capacity and self-induction. Rhigi accomplished this by coating a plate of glass with tin foil and making with a diamond a thin cut in the tin foil. Minute sparks passed across this cut when the detector was in electrical tune with the little spheres in the oil.

With this apparatus, experiments on electrical waves are brought within the range of almost any experimenter, and with it Rhigi has performed by means of electrical waves almost all the ordinary optical experiments, such as refraction of waves by prisms of pitch or other light opaque insulating substances; interference of waves; reflection of waves, from the focus of one cylindrical or spherical mirror to the focus of another; and polarization of electric waves.

The experiment of polarization has been shown by a block of wood in the following manner: In Fig. 50

* Lebedew, Ann. der Physik. und Chemie, vol. lvi, 1895.

are represented the two parabolic mirrors used by Hertz to show reflection of electro-magnetic waves, A being the transmitter and B the receiver. B is turned so that its focal line is at right angles to that of A. In this position no sparks are observable in the little resonator in the focal line of B. When a block of wood, however, is placed in the line between the exciter or oscillator of A and the resonator of B, in certain positions of the grain of the wood sparks begin to appear in the resonator. The wood allows the waves to pass through in certain positions of its fibres, and shuts them out in other positions. In other words, it polarizes the waves, just as certain substances of different molecular aggregations in different directions polarize light waves. Lebedew has used little spherical mirrors instead of parabolic mirrors, and has reduced the dimensions of the apparatus to almost the size of optical apparatus for measuring wave lengths of light. Instead of the res-

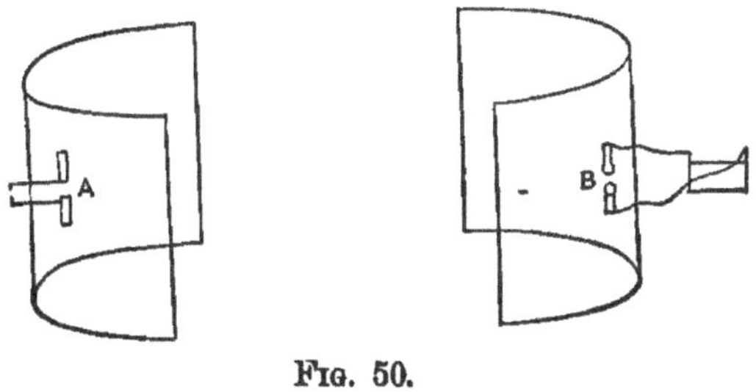

FIG. 50.

onator of Rhigi, he employs for detector of waves in the air a thermal junction. Thus we have now three forms of electrical eye, so to speak: the micrometer spark gap used by Hertz (Fig. 51, M); the plate of glass with the extremely fine diamond cut across its tin-foil-coated surface N; and the thermal junction, first used by Klemencic O. With the spark gap we detect the waves by the light of the sparks; with the thermal junction we detect the waves surging between A and B by the heat developed in the very fine junction J. On account of the self-induction, or electrical inertia of the galvanometer circuit, the waves do not pass along this, and their oscillations are confined between A and B.

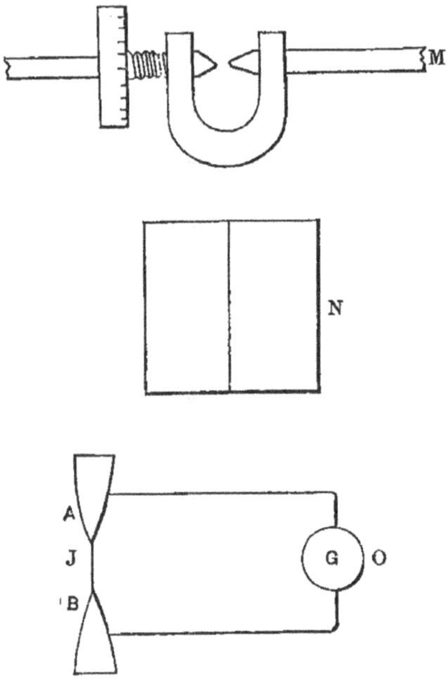

Fig. 51.

The electro-magnetic waves pass through brick walls unaffected by them. They are probably not absorbed by fog, and therefore if we could detect such waves at a distance of a thousand feet we should have a method of signalling through a dense fog, and possibly a method of preventing collisions at sea. The most powerful electric light can not penetrate a dense fog this distance. Unfortunately, at present we can not detect the electro-magnetic waves more than one hundred feet from their source. To do this we are obliged to employ the large parabolic mirrors of Hertz, and a comparatively small spark at the focus of one mirror and a Rhigi detector at the focus of the receiving mirror. Very powerful sparks do not seem to work as well as comparatively feeble ones. I have had constructed parabolic mirrors nearly six feet high and six feet across. Such mirrors are twice the size of those employed by Hertz, but I have not hitherto been able to greatly extend the distance at which the electrical waves become too feeble to be distinguished. Lord Rayleigh has shown that with long waves of sound a plane surface is as good a transmitter or reflector as a curved surface. It is probable, also, that with very powerful sparks a plane mirror would answer better than a parabolic mirror or a spherical mirror, for it is impossible to produce a very powerful electric spark with a very rapid rate of oscillation, on account of the large amount of capacity necessary. The waves, therefore, must be several feet in length. With a flat mirror we should avoid the diffraction or bending of the wave about the edges of the curved mirror. The principal difficulty, however, of extending the range of electro-magnetic waves is in their rapid rate of decay, or damping, as it is termed. The light of the electric spark can be detected by the eye much farther than the electro-magnetic waves it sends forth. Yet this light is caused by the electric waves. The electro-magnetic waves have decreased so greatly in amplitude when they have reached the position of the observer's eye that they have not sufficient energy to start waves in any detector with which we are now acquainted. They can, however, disturb the ether so as to give the sensation of light.

A very important criticism on the present methods of measuring the reflection of electric waves and their supposed interferences in air has been made by Bjerknes. He points out that the amplitude or height of the electric waves change as they progress away from their source, while the amplitude of plane light waves remain unchanged. The electric waves can be represented by curve similar to the curves in Fig. 52, in which the successive crests diminish or increase in height, due to conditions of the oscillating source of electricity, while the regular amplitudes of the light waves can be represented by the regularly undulating curve of Fig. 15, page 143. Furthermore, we study optical phenomena by apparatus which is inert, so to speak—that is, the apparatus does not produce waves of light itself, which complicate the observations. Thus we observe the spectrum by means of a telescope; we study the interference bands of light produced, for instance, by light in passing through a very narrow slit by means of the eye. Now the telescope and the eye do not produce waves of light which interfere with the observations. In the study, however, of electro-magnetic waves in air our only method at present is by the employment of electric circuits in which a spark gap is placed. Now the electric wave sent out from the spark gap of the exploring circuit also has a varying amplitude, and therefore when we move, for instance, our resonator—that is, the little circuit inclosing a spark gap—to and fro in front of a wall which is reflecting electric waves, the interferences we perceive may be merely the interferences between the decreasing amplitudes of

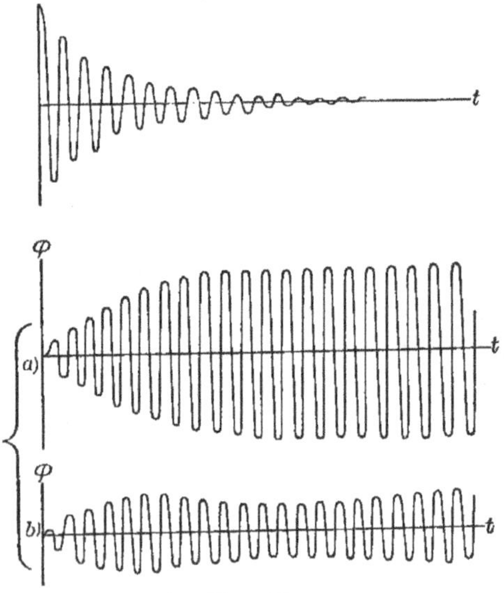

FIG. 52.

the wave reflected from the wall and the changing amplitude of the wave in the resonator.*

If we look at the incandescent filament of a distant Edison lamp through a narrow slit cut in card, or if we gaze at it along the edge of the card, we perceive interference bands of light, consisting of bright and dark spaces. The eye may be called an indifferent instrument; it passively accepts the phenomena. The waves of light interfere with each other on the retina. When the crest of one wave coincides with the trough of another, no impression is made on the retina; we see a dark space. When the crests of the waves coincide, we are conscious of a bright space. If now the retina oscillated, so to speak, or if, on being excited, it sent out waves also, it is evident that we should be conscious of very complicated phenomena of interference. The waves in our eyes might annul the waves coming from the outer object so that we might not be conscious of the outer light.

The experiments on electrical resonance lead us to believe strongly in the existence of a periodic electrical disturbance in some medium filling space, which is propagated with the velocity of light. The velocity of electricity in this medium, according to Maxwell, is the same as the velocity of light. This is evidently a very important point to establish in regard to Maxwell's great generalization, and in connection with Mr. Duane, one of my graduate students, I have carefully measured the velocity of electric waves by the following method, which is free from estimations of electrical dimensions of the circuit, and by which the wave lengths and the time were measured directly on the same circuit along

* V. Bjerknes, Annalen Physik und Chemie, No. 1, 1895.

which the wave motion was propagated. This method of procedure is of great importance; for we have seen that the early methods of measurement of the velocity of electricity were vitiated, so to speak, by the conditions of the apparatus which was used for the measurement, and later calculations depend upon estimations of self-induction and of the time of the periodic motion of a circuit, which is not the same as that along which the motion takes place. In short, our measure is a direct measure of the speed of the electric waves in the ether near the surface of the wires we employed.

The arrangement and dimensions of the apparatus finally adopted were as follows (see Fig. 45, page 246):

Two metallic plates, *a* and *b*, 30 × 30 centimetres, placed in vertical planes, formed the primary condenser. The dielectric between them consisted of the best French plate glass obtainable, and was two centimetres thick. Outside the plates *a* and *b*, and separated from them by a hard-rubber dielectric, 1·8 centimetre thick were the secondary plates 26 × 26 centimetres. The primary and secondary circuits were joined to the condenser plates as indicated in the figure. The primary circuit lay in the horizontal plane passing through the centres of condenser plates, and consisted of copper wires, 0·34 centimetre in diameter. In order to control the period of oscillation of the primary circuit, the portion B D, containing a spark gap with spherical terminals, was made to slide along parallel to itself. The distance between the straight portions A B and C D was 40 centimetres, and the lengths of A B and C D finally chosen for best resonance were 85 centimetres. Most of the secondary circuit lay in a horizontal plane 15 centimetres above that of the primary. The lengths G E and H F, however, were bent down and fastened to the middle points, G and H, of the secondary plates. The circuit consisted of copper wire (diameter 0·215 centimetre), and its total length from G through I to H was 5,860 centimetres. At I was a spark gap with pointed terminals. With this apparatus we succeeded in producing a very regular wave formation, as indicated by the bolometer. There was a node at I, and another about 40 centimetres to the right of E and F.

The images of the secondary spark were thrown on a sensitive plate by means of a rotating mirror. The dots obtained represent discharges from the negative terminals only, the positive discharges not being brilliant enough to affect the plate. The distance between successive dots was the distance on the plate through which the image of the spark gap moved during the time of a complete oscillation. Hence by determining the speed of the mirror and measuring the distances from the mirror to the plate the time of oscillation could be calculated. To measure the sparks, we used a sharp pointer moved at the end of a micrometer screw, under a magnifying glass of low power. The instrument was originally intended for microscopic measurements, and was very accurately constructed. The rotating mirror was driven by an electric motor by means of a current from a storage battery of extremely constant voltage. To give great steadiness, a heavy fly wheel was attached to the axis of the mirror. The speed of the mirror was determined to within about one part in five hundred by means of an electric chronograph.

It appears from the best results that we have obtained that the velocity of short electric waves travelling along two parallel wires differs from the velocity of light by less than two per cent of its value. It has been shown theoretically that the velocity of such waves travelling along a single wire should be the velocity of light approximately. These results, therefore, in a certain sense, confirm the theory, to an accuracy within their probable error.

Theoretically, too, the velocity should be approximately equal to the ratio between the two systems of electrical units. The average of the best measurements of this ratio is 3·001, which is nearer the average velocity obtained for electric waves than the velocity of light.

We have established, I believe, beyond reasonable doubt, that the waves of electricity travel with the velocity of light, for the waves on the wires we employed must have been proved to reside entirely on the surface —that is, on the boundary, so to speak, of the medium pervading the space about the wires. The formation of stationary electric waves which are propagated with the velocity of light is the best evidence we now have of the truth of Maxwell's great generalization.

We have said that Joseph Henry showed at an early date that the discharge of a Leyden jar is oscillatory. Our present knowledge of electric waves is largely due to a realizing sense of the importance of the observations of Henry. One will find in his published paper the following remarkable conclusion, which can be regarded almost as a prophecy:

"In extending the researches relative to this part of the investigations, a remarkable result was obtained in regard to the distance at which induction effects are produced by a very small quantity of electricity. A single spark from the prime conductor of a machine of about an inch long, thrown on to the end of a circuit of wire in an upper room, produced an induction sufficiently powerful to magnetize needles in a parallel circuit of iron placed in the cellar beneath, at a perpendicular distance of 30 feet, with two floors and ceilings, each 14 inches thick, intervening. The author is disposed to adopt the hypothesis of an electrical plenum, and from the foregoing experiments it would appear that a single spark is sufficient to disturb perceptibly the electricity of space throughout at least a cube of 400,000 feet of capacity. And when it is considered that the magnetism of the needle is the result of the difference of two actions, it may be further inferred that the diffusion of motion in this case is almost comparable with that of a spark from a flint and steel in the case of light." *

---

* Scientific Writings of Joseph Henry, vol. i, p. 202, Smithsonian Institution, Washington.

CHAPTER XIX.

THE ELECTRO-MAGNETIC THEORY OF LIGHT AND THE ETHER.

The various phenomena of the action between magnets—the induction phenomena between neighbouring circuits, a current of induction rising in one circuit whenever an electric current is started in a neighbouring circuit, and thus manifesting energy—lead us to believe that the energy has been transferred from the exciting circuit across space by means of some medium filling that space. The energy has disappeared from the exciting circuit, and has reappeared in the induction circuit. It must have existed during the time of its disappearance and reappearance in the intervening space. We are therefore forced to believe in some medium which serves to convey this energy.

The old fluid theories implied that when a body was electrified it had something upon it which was called electricity. According to the modern views, we regard the ether around the body as charged with energy which is the result of the work we have done in charging the body. This energy in the ether is the energy of motion. There is a state of strain in the ether which we term a polarized condition. Around a positively charged body this polarization has a certain direction and a certain amount. With a negatively charged body this polarization is in an opposite direction. It is suggested that these polarizations may be like right-handed and left-handed rotations or twists. When we electrify a conductor we store up energy around the conductor in the ether. The work we do is spent in changing the state of the medium. When a body is discharged, the medium returns to its original state, and the energy is dissipated as heat in the electric spark or as heat in the conductor. The electric current is therefore the manifestation of energy in the ether along the wire through which the current appears to flow. According to Poynting's theory, we have seen that the electric energy produced by a battery or a dynamo does not flow along the wire—for instance, the overhead wire of an electric railroad—but it produces this strained condition in the ether, and the ether relieves itself along the wire. What we call the flow of the electrical current is therefore not in the same direction as the flow of electrical energy.

When a Leyden jar is discharged the knobs of the jar become alternately positive and negative. The medium around the jar is therefore polarized alternately in opposite directions. This polarization starts from the knob and spreads through space, at each point of which there are to-and-fro motions, and waves of opposite polarizations are sent through the medium, carrying the energy which had been stored up in the Leyden jar. There is a periodic or to-and-fro movement in the ether, and if we could make a Leyden jar of molecular dimensions charge it and discharge it, we could produce a periodic movement in the ether which is analogous to that which occurs in the propagation of light. Maxwell's electro-magnetic theory of light supposes that the periodic motions which constitute light are of the same nature as those which arise when the positive and negative conditions of the ether are rapidly alternated in the case of the discharge of a Leyden jar. Light, heat, and electricity are therefore manifestations of electro-magnetic waves which come to us from the sun.

In our study of electric waves we have used their light manifestations in order to trace their phenomena. Thus, when a Leyden jar is discharged through a great circle of wire properly placed in a room, we can send electro-magnetic waves through brick walls and detect them in neighbouring rooms by the sparks that are excited in a similar circle of wire connected to another similar Leyden jar. By photographing the latter spark, we can say, in popular language, that we have photographed by means of waves that have passed through a brick wall. We shall see later that it is possible to photograph by means of electric waves which have passed through opaque metallic screens, which cut off entirely the light rays so considered.

Our eyes can see an electric spark very much farther than we can detect the electro-magnetic waves by any other instrument than the eye. The eye really detects them in the form of light. It is true also that we can not detect the heat waves sent out by the spark so far as we can perceive the electric waves. The heat waves are nearer in length to the electric waves which we can detect than the light waves. The most delicate thermometer will not show any indication of heat at a distance of ten feet from a powerful spark if we prevent the electro-magnetic waves from surging in its mass, and if we depend upon the direct radiation of heat through the ether.

We are thus certainly as advantageously placed at present in regard to measuring electro-magnetic waves generated by a spark as we are in regard to measuring heat waves which accompany them. We can, however, measure the long waves of heat which come to us from the sun, yet we can not detect the long electro-magnetic waves. Prof. J. J. Thomson has shown that if an electrical charge on a sphere is disturbed in any sudden way, it can oscillate to and fro in the time taken by light to travel a certain number of times the diameter of the sphere, depending on the wave length of the electric wave. Prof. O. J. Lodge, in quoting this calculation, remarks: "An electrostatic charge on the whole earth would surge to and fro seventeen times a second. On the sun an electric swing lasts six and a half seconds. Such a swing as this would emit waves 19×10^6 kilometres or 1,200,000 miles long, which, travelling with the velocity of light, could easily disturb magnetic needles and produce auroral effects, just as smaller waves produce sparks in gilt wall paper (Rhigi's resonators), or as the still smaller waves of Hertz produce sparks in his little resonators, or, once more, as the waves emitted by electrostatically charged vibrating atoms excite corresponding vibrations in our retina." *

With our present methods of studying electric waves we are compelled to use electric sparks which do not succeed each other continuously. The electric waves from one spark are thousands of miles on their way before another spark occurs. The dying out of the waves, or what we have called their damping, is due to the rapid radiation of the energy of the spark into space. Hertz has calculated this loss of energy in the case of a small spark from two spheres of fifteen centimetres radius (about six inches) so arranged that they sent out waves four hundred and eighty centimetres long (one

* Lightning Conductors and Lightning Guards, p. 260.

hundred and ninety-two inches). He estimates that such an oscillator giving waves of this length must be supplied with energy at the rate of twenty-two horse power per second if the intensity is to be kept constant. "At a distance of twelve metres (thirty-nine and a half feet) from the spheres or oscillator the intensity of the radiation is equal to the intensity of the solar radiation at the surface of the earth." *

From this calculation we see what an enormous amount of energy is radiated in a stroke of lightning. This consideration of the rapid damping of electric waves sent out by electric sparks is interesting from the point of view of the endeavour to obtain light by rapid electric oscillations.

The hypothesis of an ether filling all space, a medium by means of which light, heat, and electricity are transmitted to the earth from the sun, seems untenable to many minds, largely because we are obliged to attribute to this medium extraordinary qualities of rigidity and elasticity, and to regard it of such extreme tenuity that it escapes detection by direct measurement; for it does not impede the motion of bodies.

If one critically examines the advance that has taken place in the applications of electricity, one can see that even the practical man's chief consideration is in regard to the medium around his electro-magnets and his currents. In designing dynamos and motors for the transmission of power, one of the main principles to be borne in mind is that movable parts of iron or copper will arrange themselves in a magnetic field so as to diminish as far as possible the magnetic resistance of such a field. By magnetic resistance we mean the dif-

* Preston, Theory of Light, p. 444.

ficulty different media oppose to the establishment of magnetic lines of force. Thus an iron nail will set itself along the lines of force, so that they may pass through it rather than be spread through the medium outside the iron. A copper cent will tend to present its edge to the magnetic pole, for this position offers the least resistance in the field to the establishment of the lines of force or the stress in the medium.

This great law, that bodies arrange themselves so as to diminish the electrical distance of the field in which they are placed, enables us to detect rapid to-and-fro currents which our compasses and galvanometers are utterly unable to show us. If, for instance, we should put a compass near a telephone through which we speak, no movement will be perceived of the needle of the most delicate compass; if, however, we suspend a piece of soft iron in front of a telephone magnet, so that it makes a certain angle with the axis of the magnet, and provide it with a tiny mirror, we shall find that when we speak through the telephonic circuit the little bar of iron will move so as to place itself along the direction of the fluctuating lines of magnetic force. It thus betrays the distance of to-and-fro electric currents in electro-magnetic coils which otherwise might be uususpected. The stress does not reside in the air, for the magnetic lines stream through the best vacuum we can produce. The stress is in the ether.

A similar law holds in regard to lines of electric force which exist about bodies charged with electricity. If we should hang up in a room a charged pith ball, these electrostatic lines extend from the pith ball to the walls of the room; they indicate a stress or strain in the medium. All the particles of dust in the room tend to place themselves so as to diminish the resistance of the medium to the establishment of these electrostatic lines. If the length, for instance, of a human hair is in the direction of one of these lines of force, it will not indicate any rapid change in electrification of the pith ball; but if it is not in such a direction it will make manifest by its movement what would not be otherwise revealed. The electrical engineers use in their calculations a term which is called the permeability of the medium for magnetic lines of flow.

The phenomena of electro-magnetism, then, compel us to assume the existence of a medium through which and by means of which the electrical energy is transmitted. In the case of electro-magnetism, we observe that certain lines of force are established in the space around magnets, and around wires carrying currents. While these lines are being established, work must be done on the medium, and energy is stored up in the medium. When the lines of force are withdrawn from the medium, the energy that was stored up now rushes into the wire of the circuit of the electro-magnet and is dissipated as heat.

The phenomenon of self-induction manifests this storing of energy in the medium; and it is only within a few years comparatively that we have obtained a clear conception of what we now call self-induction or inductance. In the subject of electricity we may be said to rely upon the phenomenon of induction to prove the necessity of a medium between the pith-ball magnets or electric circuits which manifest the varied phenomena of induction. Since the phenomena of light seem to require the existence of a medium as well as the phenomena of electricity, a careful search has been made to establish a relation between light and electromagnetism by means of experiment. To establish such a relation we must discover some phenomenon of light which is affected by phenomena of magnetism. Mrs. Somerville endeavoured to magnetize a needle by letting a beam of light fall upon it, and apparently succeeded. Maxwell remarks in regard to this noted experiment that "we must remember that the distinction between magnetic north and south is a mere matter of direction, and would be at once reversed if we reverse certain conventions about the use of mathematical signs. There is nothing in magnetism analogous to the phenomena of electrolysis which enable us to distinguish positive from negative electricity, by observing that oxygen appears at one pole of a cell and hydrogen at the other. Hence we must not expect that if we make light fall on one end of a needle, that end will become a pole of a certain name, for the two poles do not differ as light does from darkness." *

There are, however, two experiments which appear to establish a certain experimental relation between light phenomena and magnetic phenomena. Faraday, in seeking such a relation, discovered that if a beam of polarized light (see p. 80) is sent through a parallelopipedon of dense glass placed in the core of a powerful electro-magnet, and if it is examined by an analyzer (see p. 80), it is found that the plane of polarization is rotated through a certain angle. One therefore has to turn the analyzer around its axis to shut off the light which is now transmitted through the two Nicols. The direction in which one must turn the analyzing Nicol depends upon the direction of the current which excites the electro-magnet. Here is evidently a relation between light and electro-magnetism. The mag-

---

* Maxwell's Electricity and Magnetism, vol. ii, p. 410.

netic force through the glass coincides with the direction of the ray of light; and the effect of this magnetic force is to turn the plane of polarization around the direction of the ray as an axis. This effect has been noticed in a number of substances besides glass, and the amount of the turning depends upon the nature of the substance. It has also been discovered that if a beam of polarized light is allowed to fall on the surface of the iron core of a powerful electro-magnet its plane of polarization is rotated.

After a careful consideration of phenomena of this nature, Maxwell concludes that "there is nothing in the magnetic phenomena which corresponds to wave length and wave propagation in the optical phenomena. A medium in which a constant magnetic force is acting is not, in consequence of that force, filled with waves travelling in one direction, as when light is propagated through it. The only resemblance between the optical and the magnetic phenomenon is that at each point of the medium something exists of the nature of an angular velocity about an axis in the direction of the magnetic force." The consideration of the rotation of the plane of polarization of light by magnetic force led Maxwell to a theory of magnetism which is called the hypothesis of molecular vortices. Since there is good evidence for the belief that there is some kind of rotation going on in the magnetic field, Maxwell investigated the condition of motion which exists when a great number of very small portions of matter rotate on their own axes, these axes being parallel to the direction of the magnetic force. The motion of these vortices does not sensibly affect the visible motions of large bodies, but it can be supposed to affect the periodic motion of the medium which constitutes the phenomena we call light. According to this theory, the displacements of the ether will produce a disturbance of the vortices, and this disturbance of the vortices can be supposed to react on the ether, and in this way can affect the propagation of light. The mathematical discussion of this theory leads to the following conclusions:

1. Magnetic force is the effect of the centrifugal force of the vortices.

2. Electro-magnetic induction of currents is the effect of the forces called into play when the velocity of the vortices is changing.

3. Electro-motive force arises from the stress on the connecting mechanism.

4. Electric displacement arises from the elastic yielding of the connecting mechanism.*

While thus the varied manifestations of the transformations of energy which we witness in the subject of electricity apparently compel us to assume the existence of an ether pervading all space, there is no hypothesis of physics which seems more arbitrary than that of the ether to the cosmological philosopher.

The Marquis of Salisbury, in his address before the British Association, Oxford, 1894, says of the ether:

"It may be described as a half-discovered entity. I dare not use any less pedantic word than entity to designate it, for it would be a great exaggeration of our knowledge if I were to speak of it as a body or even a substance. When, nearly a century ago, Young and Fresnel discovered that the motions of an incandescent particle were conveyed to our eyes by undulations, it followed that between our eyes and the particle there must be something to undulate. In order to furnish

* Maxwell's Electricity and Magnetism, vol. ii, § 831.

that something, the notion of the ether was conceived, and for more than two generations the main, if not the only, function of the word ether has been to furnish a nominative case to the verb 'to undulate.' Lately our conception of this entity has received a notable extension. One of the most brilliant of the services which Prof. Maxwell has rendered to science has been the discovery that the figure which expressed the velocity of light also expressed the multiplier required to change the measure of the static or passive electricity into that of dynamic or active electricity. The interpretation reasonably affixed to this discovery is that, as light and the electric impulse move approximately at the same rate through space, it is probable that the undulations which convey them are undulations of the same medium. And as induced electricity penetrates through everything, or nearly everything, it follows that the ether through which its undulations are propagated must pervade all space, whether empty or full, whether occupied by opaque matter or transparent matter, or by no matter at all. The attractive experiments by which the late Prof. Hertz illustrated the electric vibrations of the ether will only be alluded to by me, in order that I may express the regret deeply and generally felt that death should have terminated prematurely the scientific career which had begun with such brilliant promise and such fruitful achievements. But the mystery of the ether, though it has been made more fascinating by these discoveries, remains even more inscrutable than before. Of this all-pervading entity we know absolutely nothing except this one fact, that it can be made to undulate. Whether outside the influence of matter on the motion of its waves ether has any effect on matter or matter upon it, is absolutely unknown. And even its solitary function of undulating ether performs in an abnormal fashion which has caused infinite perplexity. All fluids that we know transmit any blow they have received by waves which undulate backward and forward in the path of their own advance. The ether undulates athwart the path of the wave's advance. The genius of Lord Kelvin has recently discovered what he terms a labile state of equilibrium, in which a fluid that is infinite in its extent may exist, and may undulate in this eccentric fashion without outraging the laws of mathematics. . . . In any case it leaves our knowledge of the ether in a very rudimentary condition. It has no known qualities except one, and that quality is in the highest degree anomalous and inscrutable. The extended conception which enables us to recognize ethereal waves in the vibrations of electricity has added infinite attraction to the study of those waves, but it carries its own difficulties with it. It is not easy to fit in the theory of electrical ether waves with the phenomena of positive and negative electricity; and as to the true significance and cause of those counteracting and complementary forces to which we give the provisional names of negative and positive, we know about as much as Franklin knew a century and a half ago."

It is true that the phenomena presented by the action of two electrified pith balls are still full of mystery, but not more so than many phenomena which we perceive daily and do not closely examine. One will find it very difficult, for instance, to explain what takes place in the ether when we light a candle. We can trace the waves of light and apply our mathematical processes when the waves are investigated at some distance from the candle, and when they become what are called plane waves; but when we endeavour to understand what takes place at a point of light when the spherical waves receive, so to speak, their primal impulse, we are certainly as much puzzled as we are to account for the action of two electrified pith balls. This certainly can be maintained, that the hypothesis of a medium and the theory of action from point to point in the medium—the theory of what the Germans call *Nahekräfte*, in opposition to that of *Fernkräfte*, or action at a distance—has enormously increased our just conceptions of the transformations of electrical energy.

The hypothesis of action at a distance seems at variance with the modern ideas of the continuity of matter. It was naturally suggested by the attraction between the moon and the earth and the attraction of the heavenly bodies in general with regard to each other. Perhaps nothing marks so strongly the modern attitude toward physical manifestations as the substitution for action at a distance the action in matter from particle to particle. The pole of a magnet does not attract or repel the pole of another magnet by a direct action through space which is not influenced by the matter in this space. The mutual effect of the poles depends on an action from point to point in the medium between the poles. Every magnetic pole sends out lines of force in all directions. If one magnetic pole is brought near another one, the force of attraction or repulsion between them immediately becomes manifest, and at the same time the disposition of the lines of force of one magnetic pole is altered by the lines of force of the neighbouring pole. We ordinarily say that the field of force of one pole is distorted by the entrance of the other pole.

It has also been shown that there is a rotary action of the medium near the poles of a magnet. This rotary action is from point to point in the medium surrounding the magnetic poles, and it undoubtedly is concerned in the attraction phenomenon which two pieces of magnetized steel exhibit.

Although Faraday and Maxwell advanced the study of the transformations of energy to a marked degree by the conception of action from point to point in place of the old theory of action at a distance, it seems to me that even between the ultimate atoms of matter we have what is essentially action at a distance. The distance may be infinitely small, still it is only small relatively.

The theory that all states of matter fade into each other by insensible degrees, and that what we term atoms are merely whirls in a universal medium, enables us by a forced theory to escape from the theory of action at a distance.

Prof. Emil du Bois-Reymond, in his remarkable little treatise entitled Die Sieben Welträthsel (The Seven World Mysteries or Puzzles), says: "A physical atom—that is, a body which is disappearingly small in comparison with the bodies which appeal to our senses, yet, in spite of its name, a still divisible mass to which physical properties and conditions of movement can be ascribed and of which in innumerable numbers larger masses are composed—is an extremely useful fiction, especially in chemistry and in the mechanical theory of gases. The tendency, however, in mathematical physics is to shun the hypothesis of atoms and to replace the discrete atom by the volume element of a continuous medium."

"Newton endeavoured to account for gravitation by differences of pressure in an ether," but he did not publish his theory "because he was not able from experiment and observation to give a satisfactory account of this medium and the manner of its operation in producing the chief phenomena of nature." The phenomena of light also demand the hypothesis of a medium. No one has better summed up the arguments for the existence of an ether than Maxwell, and I can not do better than give his principal arguments as follows: *

We can prove by means of the phenomena of interference that light is not a substance. If light from a candle is divided into two parts which are made to unite after traversing two different paths and to fall on a screen, and if either half of the beam is shut off by a screen, the other half will still illuminate it. If, however, we examine the light on the screen when both portions of the beam are allowed to fall together on it, we perceive certain dark bands crossing the screen. At these dark points the waves of light have interfered with each other. The trough of one wave coincides with the crest of another wave and darkness results; or, in ordinary language, we say that one portion of light has destroyed another. If light were a substance, one portion of it added to another portion could not make it cease to be evident to our senses. The only explanation of the interference of two rays of light that is possible is this, that it arises from a periodic movement in a medium.

We conclude that light is a process, and not a substance. The medium is capable of transmitting energy, as is seen from the phenomena of electricity and of light. This energy is not transmitted instantly from the radiating body to the absorbing body, but exists

---

* Encyclopædia Britannica, Ether.

for a time in the medium. Let us therefore examine the necessary physical properties of this ether: The coefficient of rigidity of ether $= 842{\cdot}8$; the density of the ether $= 9{\cdot}36 \times 10^{-19}$; the coefficient of rigidity of steel is about 8×10^{11}; and that of glass, $2{\cdot}4 \times 10^{11}$. It has been computed that if the temperature of the atmosphere were everywhere 0° C., and if it were in equilibrium about the earth supposed to be at rest, its density at an infinite distance from the earth would be 3×10^{-346}, which is about 3×10^{327} times less than the estimated density of the ether. In the regions of interplanetary space the density of the ether is therefore very great compared with that of the attenuated atmosphere of interplanetary space.

Air can not transmit transverse vibrations, and the normal vibrations which the air transmits in the case of sound travel about a million times slower than light. We must suppose that the medium (ether) is different from the transparent media known to us. The energy of vibration of gross particles is very much less than that of the ether; otherwise a much greater proportion of incident light would be reflected when a ray passes from vacuum to glass or from glass to a vacuum than we find.

Faraday says: "For my own part, considering the relation of a vacuum to magnetic force and the general character of magnetic phenomena external to the magnet, I am much more inclined to the notion that in the transmission of the force there is such an action external to the magnet, than that the effects are merely attraction and repulsion at a distance. Such an action may be a function of the ether, for it is not unlikely that, if there be an ether, it should have other uses than simply the conveyance of radiation."

The following are the objections to the undulatory theory:

1. The theory indicates the possibility of undulations of vibrations normal to the surface of the waves. To account for no optical effects we have to assume the incompressibility of ether.

2. The phenomena of reflection are best explained on the hypothesis that the vibrations are perpendicular to the plane of polarization; those of double refraction require us to assume that the vibrations are in that plane.

3. In order to account for the fact that in a doubly refracting crystal the velocity of the rays in any principal plane and polarized in that plane is the same, we must assume certain highly artificial relations among the coefficients of elasticity.

Maxwell, in thus stating these objections, concludes with the following defence of his theory: "The electromagnetic theory satisfies all these by the single hypothesis that the electric displacement is perpendicular to the plane of polarization. No normal displacement can exist; and in doubly refracting crystals the specific dielectric capacity for each principal axis is assumed to be equal to the square of the index of refraction of a ray perpendicular to that axis and polarized in a plane perpendicular to that axis. Boltzman has found that these relations are approximately true in the case of sulphur, a body having three unequal axes.

"Ether transmits transverse vibrations to very great distances without sensible loss of energy by dissipation. A molecular medium, moving under such conditions that a group of molecules once near each other remain near each other during the whole motion, may be capable of transmitting vibrations without much dissipation of energy; but if the motion is such that the groups of molecules are not merely slightly altered in configuration, but entirely broken up so that the component molecules pass into new types of grouping, then in the passage from one type to another the energy of regular vibrations will be frittered away into heat. We can not therefore suppose the ether like a gas.

"The ether, though homogeneous and continuous, may be, as regards its density, rendered heterogeneous by motion (Hypotheses of Vortex Molecules, Lord Kelvin). Magnetic influence on light indicates a rotational motion of the media when magnetized. This motion does not imply a dissipation of energy.

"No theory of the ether will account for such a system of molecular vortices being maintained for an indefinite time without their energy being frittered into heat."

In support of this great generalization of Maxwell, it can be said that none of the investigations of the great army of physical investigators since the death of Maxwell tend to disprove, but rather to prove, the truth of his generalizations. It rests upon the hypothesis of the ether; and, as we have seen, the various transformations of energy require for their explanation the existence of a medium. Among the most interesting investigations which tend to prove the existence of the ether are those in regard to the cathode rays. The word cathode is from the Greek *κατα*, down, and ὁδος, a way. It is applied to the negative terminal of a battery or to the negative terminal of a Ruhmkorff coil or transformer. It can also be applied to the outside coating of a Leyden jar, the interior of which is charged positively, the outside being thus charged negatively. The name arose from the old and earlier supposition that electricity flowed from the positive pole down to the negative pole.

Now, if we imbed the wires connecting the cathode and its reverse, the anode, of a Ruhmkorff coil, in a little glass vessel and exhaust the vessel of air, when we excite the coil and when we reach a high state of exhaustion in the vessel it is filled with luminosity, and the cathode rays can be recognised as streaming out from the cathode.

A marked peculiarity of the cathode rays consists in this, that they are independent of the position of the anode, and, after emerging from the cathode, they continue in straight lines and apparently do not seek the anode. Thus, if both cathode and anode are placed one above the other at one end of an exhausted tube, the cathode rays continue to the end of the tube, and do not bend to the position of the anode.

Prof. Crookes believes that the phosphorescent effects produced by the negative rays are due to the impact of the molecules of the gas on the phosphorescent substances which emit light. It is still undecided whether the luminous appearance is due to such impacts, or whether the effect is entirely electrical. Prof. J. J. Thomson points out that the phosphorescent substance on which these cathode rays fall is subjected to very rapid periodic change of polarization, which, according to the electro-magnetic theory of light, would produce the same effect as if light fell on the phosphorescent substance, in which case we know that it would phosphoresce.

In considering the remarkable light effects produced by currents of high frequency, we perceive that we have to do with an increased activity of the molecules of a rarefied gas which is produced by the electrical energy stored up in the medium near the electrodes of our little lamps. Thus we have a bombardment of these electrodes or of the walls of the inclosing vessel; and these walls can suitably reflect or direct these molecules, as we have seen in Prof. Crookes's experiment. The cathode rays may be considered as a radiation of electric energy, which is made visible in rarefied media and which can also be detected outside such media. The visible transformation of electric energy from the electrode, termed the cathode, can be made to pass through metal walls, and can be seen outside these metal walls. This phenomenon is one which at present is attracting perhaps more attention from scientific men than any other in electricity, for in this phenomenon we see a manifestation in the medium which may lead to a better understanding of the polarization of the ether.

Before describing more minutely the interesting apparatus by which this phenomenon is studied, let us grasp the salient facts. The cathode rays can be made to pass through substances which are entirely opaque to ordinary light. We can, so to speak, see what may be termed an electrical candle through a wall, for these cathode rays can be made to pass through sheets of aluminium, and gold and silver, and many other opaque substances which cut off ordinary light entirely.

The sheets of aluminium which allow these electric-light rays to pass and cut off ordinary rays is 0·00265 millimetre (about $\frac{1}{9000}$ of an inch) thick. The rays are perfectly visible in the open air of a dark room after they have passed into it from a rarefied tube through such thin opaque metallic sheets. The rays spread out in all directions. They excite phosphorescent bodies, such as uranium glass, to a brilliant glow, and blacken photographic plates.

Lenard has described the phenomenon of cathode rays very fully, and has investigated it with an apparatus which, in its essentials, consisted of a Crookes tube (Fig. 28, page 193) provided with a window.

The window is hermetically closed by a thin aluminium sheet through which the rays pass into the outer room, and the whole apparatus is inclosed by a metallic vessel, which is connected to the ground to lead off disturbing electric charges. A quartz plate one half millimetre thick, placed between the window and a phosphorescing body, extinguishes the light. Thin plates of aluminium, gold, and silver allow the light to pass through. The light appears between such plates and the window, and also beyond the plates, while the plates themselves remain dark. The opacity of the quartz plates and the transparency of the metal plate to these rays form the marked difference between the cathode rays and the light rays. It is probable, however, that all substances in sufficiently thin sheets are transparent to the cathode rays. Soap films stretched on wire supports cut off the rays when they are thicker than 0·0012 millimetre. Aluminium plates, however, 0·027 millimetre thick allow the rays to pass.

In the case of light, one body in a layer one hundred millionth of a metre thick can be more opaque than another body a metre thick. Such enormous differences, however, do not exist in the case of the cathode rays. The ordinary atmosphere is an unfit medium for the passage of the cathode rays, for they speedily lose in it their movement in a straight line and become greatly diffused. The rays from the cathode window are diffused very much as the rays of sunlight are diffused in passing from a slit or opening into milk and water.

No heating effect of the cathode rays has been

detected. Lenard placed a delicate thermal junction in the rays which showed no heat effects, while a candle at the distance of fifty centimetres gave with the same thermal junction a marked effect.

It was noticed that electrified bodies lose their charges when the cathode rays fall upon them, or when they are placed in the neighbourhood of the window through which the rays pass. It is well known that when the exhaustion in the tubes in which the cathode rays are produced is pushed to an extreme, so that the vacuum is well-nigh perfect, the rays can no longer be produced, and, in fact, all electrical manifestations visible as light phenomena disappear. The vacuum seems to afford an infinite resistance to electricity. To test the question whether the cathode rays could travel in a vacuum, although they could not be excited in it, Lenard arranged a tube 1·50 metre long (4·92 feet), into which the cathode rays could pass from the rarefied tube in which they were generated. This long tube was then made as perfect a vacuum as modern methods permit. The pressure of the air remaining in the tube was only 0·000009 millimetre mercury, or $0{\cdot}01 \times 10^{-6}$ of an atmosphere.

The cathode rays streamed from the cathode through the window into the long cylinder which inclosed the vacuum. They were not visible until they struck a little phosphorescent screen, which could be moved along the tube to different distances by means of a magnet, there being a bit of iron on the screen. The rays travelled in straight lines in the vacuum, and could be detected at the end of the tube, 1·50 metre from the window. Lenard remarks that the ether is therefore the medium by which the rays travel and in which they manifest their peculiar phenomena. The motions in the ether must be of such extremely minute order that the size of the molecules is of relative importance. The molecules of gas muddle, so to speak, the ether, but it is noticeable that it is only the mass of the molecule which influence the phenomena.

If one could measure the velocity of the cathode rays in the ether, and observe their refraction by different media, one could connect the phenomena still more closely with ordinary light waves. It seems as if the study of such phenomena in the ether is destined to greatly increase our knowledge of the relations between the various forms of energy. Rays similar to the cathode rays could pass from the sun to the earth through the ether of space and exercise an effect in our atmosphere, although they would be invisible in the vacuum which exists in space between the sun and the earth.*

The electro-magnetic theory of light demands that a vacuum should be a nonconductor of electricity; for if it were a good conductor, it would be opaque to the electric waves, and, according to the electro-magnetic theory, no light would come to us from the sun. In experiments with the cathode rays we find that they can pass through certain conductors in thin layers. They are cut off, however, by layers of appreciable thickness.

---

* Ann. der Physik und Chemie, 51, 1894, p. 225.

CHAPTER XX.

THE X RAYS.

SINCE the writing of the previous chapter interest in the remarkable phenomena of the cathode rays has been reawakened to a marked degree by the discovery of Prof. Röntgen, who, by the use of ordinary dry plates and without the use of an aluminium window, has taken photographs through wood and through the human hand by means of what he terms the X rays, which he supposes are excited either in the glass walls of the Crookes tube or in the media outside the tube by means of the cathode rays.

We see, therefore, that the literature of the subject must be sought in the papers of Hittorf, Crookes, Hertz, Lenard, and Röntgen; and the interest in the mysterious manifestations of these invisible rays is twofold: first, in regard to the possible application of the phenomena to surgery, since the rays show a specific absorption, passing more easily through the flesh than through bones or glass or metallic particles; and, secondly, in relation to the questions whether we are dealing here with radiant matter shot forth from the negative pole or cathode or with longitudinal waves of electricity.

The term cathode, we have seen, is applied to the zinc pole or negative pole of an ordinary battery. It is that terminal of an electrical machine which glows least in the dark when the machine is excited. It is the shortest carbon in the ordinary street electric lamp. The positive carbon or anode burns away twice as fast as the negative carbon or cathode. If the electric light is formed in a high vacuum by means of a great electro-motive force, we no longer have a voltaic arc or a spark; instead of this the exhausted vessel is filled with a feeble luminosity, and a beam of bluish rays is seen to stream from the negative terminal or cathode. When these rays strike the glass walls of the vessel they excite a strong fluorescence. If the glass contains an oxide of uranium this fluorescence is yellow; if it contains an oxide of copper it is green. Röntgen supposes that this fluorescence excited by the cathode rays is connected in some way with the formation of what he terms the X rays. Now, a photograph of the bones in the hand, for instance, can be obtained by placing a sensitive plate in an ordinary photographic plate-holder, and by resting the hand on the undrawn slide in the daylight, with the palm of the hand outward and toward the cathode, and about six inches away from it; the bones of the hand are thus brought in the nearest possible position to the sensitive plate. At the time of the present writing, the breast and the abdomen of the human body present too great thickness for successful photographs, and the attempts to obtain representations of the cavity in which the brain is situated have been failures, since the rays do not show any marked difference in fleshy tissues. Nothing can be obtained in these attempts to photograph the brain but a contour of the cavity in which it is situated, and possibly a shadowy representation of a bullet which might be imbedded in the head. The method of obtaining a successful photograph of the hand shows the present limitations of the method. In order to obtain a fairly sharp shadow of a bone or of a shot, it should not be more than an inch away from the sensitive plate. The term shadow, however, is somewhat misleading. The photograph of the hand by the X rays is entirely different from one produced by resting the hand in a similar position to that above described against an uncovered sensitive plate in a dark room and then lighting a match. By the last method we should obtain a true shadow of the hand, the flesh would throw as dense a shadow as the bones, and the latter could not be detected in the general blackness. In the cathode photograph, on the other hand, a difference in absorptive power is shown: the flesh looks like a hazy film around the skeleton, and even the medulla cavities can be made out, and the varying thickness of the bones is more or less shown. This specific absorption is of great scientific interest as well as of practical importance.

Now, these X rays will penetrate several inches of wood, with varying amount of absorption, but they are almost entirely cut off by glass as thick as a window pane. They pass through thin layers of aluminium, even layers as thick as a silver ten-cent piece, while the silver coin almost entirely intercepts them.

It therefore immediately occurs to one, Why not return to Lenard's tube, provide a Crookes tube with an aluminium window, and thus save the great absorption of the glass walls of the tube? There are certain practical difficulties in the way. The aluminium must be very thin. Lenard used a window which we have seen was about one nine thousandth of an inch thick, and it was necessarily very small, in order to stand the atmospheric pressure. An aluminium window one eighth of an inch thick, or as thick as a ten-cent piece, would absorb nearly as much as the glass walls of the present forms of Crookes tubes, which are not more than one sixtieth of an inch thick. Glass vessels seem at present to be more practical than any composite form, in which aluminium is glued to a glass-supporting vessel: first, because they can be blown very thin, and in a shape strong enough to withstand the atmospheric pressure; secondly, because the occluded air can be more effectively driven off the inner walls of the vessels by heating it while it is being exhausted than it can be expelled from a vessel of any other material.

To obtain successful photographs, the exhaustion of the air must be pushed to a high degree. Moreover, a high electro-motive force is necessary. Pictures can be taken in one second of the skeleton of the human hand by means of high vacua tubes excited by high electro-motive force. Even in this bare recital of the present limits of the application of the X rays to photography we perceive great possibilities in the application of the method to the surgery of the human extremities. There is no doubt that small foreign bodies, like shot and pieces of glass, can be detected in the fleshy tissues of the hand. Certain accessible regions of the body, like the mouth, can possibly be examined by placing a sensitive film inside the mouth and the cathode outside of the cheek; and it does not seem improbable that a suitable cathode vessel can be inserted into certain abdominal regions and a photograph be obtained by placing a sensitive plate on the outside of the body. I have shown that, by employing two cathodes at the proper distance apart,* stereoscopic

* American Journal of Science, March, 1896.

representations of the bones can be obtained, and an estimate formed of the position of foreign bodies.

It seems to be now well established that the radiations characteristic of the X rays proceed from the solid body upon which the cathode rays impinge. One of the most successful forms of Crookes tubes for producing the Röntgen photographs is called a focus tube (devised in the chemical laboratory of King's College, London). It consists of a tube similar to that represented in Fig. 28. The cathode rays impinge on a thin plate of platinum constituting the anode, which is inclined forty-five degrees to the axis of the concave mirror. Prof. Elihu Thomson has devised a double-

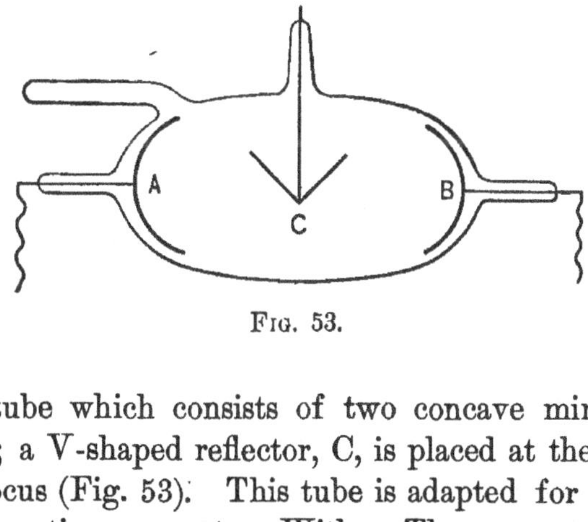

FIG. 53.

focus tube which consists of two concave mirrors, A and B; a V-shaped reflector, C, is placed at their common focus (Fig. 53). This tube is adapted for the use of alternating currents. With a Thomson and Tesla coil, practically instantaneous photographs can be taken of objects which are placed close to the photographic plate. The method of studying the effects of the X rays by means of fluorescent screens is more expeditious than that of photography. A fluorescent screen is simply a sheet of pasteboard covered with a fluorescent substance. Edison has discovered that crystallized tungstate of calcium is highly fluorescent. A pasteboard covered with this substance forms the closed end of a box into which one looks, the hand or limb being pressed against the outside of the pasteboard screen, the fluorescent substance being on the side at which one looks—that is, inside the box. By means of such a fluoroscope one can see the shadow of one's hand after the X rays have passed through several doors, and at a distance of at least fifteen feet from the Crookes tube. Sensitive photographic plates are fogged through brick walls a foot thick.

Let us now return to some of the interesting scientific questions which have arisen in regard to this apparently new manifestation of the cathode rays. In the first place, they are apparently not refracted by paraffin, vulcanite, or wood, or by any substance which is penetrated by them. To test this, I employed a double-convex lens of wood, and also a double-concave lens of the same material. I placed two copper rings in the concavity of the double-concave lens of wood, and also a similar copper ring outside the lens at the same height from the sensitive plate as one of the rings which rested on the wood of the lens. I also placed a ring on the double-convex lens, and employed two cathodes to obtain two shadows from different positions. The thickness of the wooden lenses varied from half an inch to three quarters of an inch. The images obtained through the wood of the lenses were not distorted or changed in figure in any way by the wood, and therefore no refraction could be observed by this method. On account of the quick diffusibility of the rays, no accurate method of determining a possible index of refraction seems possible. If the photographic effect is due to longitudinal waves in the ether, and if these waves travel with great velocity, no refraction would probably be observed. Maxwell's electro-magnetic theory of light supposes that only transverse waves are set up in the ether, and no longitudinal waves exist. On the other hand, Helmholtz's electro-magnetic theory of light postulates longitudinal waves as well as transverse waves. The longitudinal waves travel with an infinite velocity. Is it therefore possible that the X waves are the longitudinal waves of Helmholtz's theory? Our apparent inability to refract the rays lends colour to this hypothesis. Röntgen, in the preliminary account of his experiments, intimates that the phenomena may be due to longitudinal waves; and in a late article by Jaumann, entitled Longitudinal Light,* Maxwell's electro-magnetic equations are modified so as to embrace the phenomenon of cathode rays; and the author shows that even Maxwell's theory can, under certain conditions, give a longitudinal wave.

The cathode rays can be deflected by a magnet, and it is said that the X rays can not. It must be borne in mind, however, that when the cathode rays are widely divergent it is difficult to deflect them by a magnet; the stream density, so to speak, is too feeble. The X rays, therefore, may be only cathode rays modified by passing through the glass vessel; and the stream of rays may be of too feeble a character to be influenced by a magnet—that is, they may be still cathode rays. The want of refractive power and the want of magnetic action have not been fully established. The electrostatic lines of force go out from a charged conductor at right angles to the surface of the conductor. I have had constructed a Crookes tube with two parallel terminals

* Annalen der Physik und Chemie, No. 1, 1896.

of aluminium. The fluorescence in the walls of the vessel, when it was exhausted, showed that the cathode rays went out from every element of the cathode at right angles to it. By bending the cathode into an arc of a circle the cathode beams travelled over the surface of the vessel, forming zones of light the centres of which were in the bent wire. Is it not possible that by the electrostatic action the few molecules of air left in the high vacua are shot off with great velocity and bombard the walls of the vessel, and thus give rise to the fluorescent light, and also to an agitation of the molecules of matter outside the vessel? This may be called the molecular view of the phenomenon. I confess it is difficult to see why the molecular agitation is stopped by a thin sheet of glass and not by an inch of wood. It is certain that a few molecules must be left in the high vacua, for the cathode rays can not be formed in a perfect vacuum.

It is also true that it is useless to attempt to obtain photographs in any reasonable time from tubes which do not show a strongly marked cathode beam, or from tubes which on reversing the electric current through them do not show a marked difference between the light at the cathode and that at the anode. In poorly exhausted tubes one can perceive a faint appearance of a cathode beam, which is lost at a short distance from the cathode, as if the molecules which are shot off meet with such a crowd of more slowly moving ones that their energy is soon lost, and the cathode beam is quickly diffused like a beam of sunlight passing into milk and water. Thus the beam of cathode or X rays emerging from the glass vessel into the air is soon no longer conical in form. The sides of the cone of rays are no longer straight; they are curved, as if the generatrix of the cone were a curved line instead of a straight line, and the beam is soon lost in a turbid medium. One can imagine a stream of projectiles being similarly dispersed in striving to pass into a region of sluggishly moving shot. This molecular view of the phenomenon seems at first sight to be a more tangible one than the longitudinal wave theory. Yet the amount of energy required by any corpuscular theory would seem to be enormous. It is possible, too, that the impact of the molecules on the aluminium window of Lenard, or on the glass sides of the vessel, may serve to start ripples, so to speak, in the ether, which are propagated with the velocity of light.

The Röntgen phenomenon seems to be a manifestation of cathode rays brought to light and endowed with great practical interest by its application to dry-plate photography. When we return to the classical investigation of Lenard mentioned in the last chapter, we are impressed by his apparently crucial experiment which he describes in regard to the existence of an ether or medium. Energy passed into the vacuum he formed, and could be detected from point to point. We can conceive of its passing through the ether in the tube by a wave motion, but not by a molecular movement, for there were no molecules to move. The molecular bombardment must have stopped at the aluminium window, and the resulting energy may have been propagated by ripples in the ether. This experiment of Lenard seems to me the most interesting one in the subject of cathode rays. The greatest mystery, however, which envelops the subject is the action of the X rays on bodies charged with electricity. When the rays fall on, for instance, a charged pith ball, the charge disappears. Prof. J. J. Thomson and Prof.

Rhigi have found that a positive as well as a negative charge is dispelled by the X rays. The energy of the medium about the pith ball is changed to a marked degree, and in this phenomenon we seem to be brought closer to a wave theory in a medium than to a molecular theory of movement of matter.

The tendency at present is to believe that the X rays are waves of ultra-violet light of much smaller dimensions than any that have been hitherto detected. D. A. Goldhammer* strongly advocates this view. Prof. Röntgen's reasons for believing that the new radiations discovered by him were not those of ultra-violet light were as follows:

a. The X rays suffer no refraction in passing from air to water, bisulphide of carbon, aluminium, rock salt, glass, zinc, etc.

b. They are not regularly reflected by known bodies.

c. They can not be polarized by known means.

d. The density of a body apparently influences their absorption more than that of any other factor.

If the X rays are very short transverse waves of light which are too small in comparison with unevenness of highly polished substances to be regularly reflected or polarized, *b* and *c* can be explained.

When we consider also the phenomenon of anomalous refraction and dispersion the behaviour of the so-called X rays is not so remarkable. Certain substances, like fuchsin and aniline, exhibit anomalous refraction—that is, a ray of blue light may be more refracted in passing through a solution of these substances than a ray of violet light; while with substances like glass,

* Annalen der Physik und Chemie, No. 4, 1896.

which exhibit normal refraction, the violet rays are more refracted than the blue rays.

In certain cases of anomalous refraction and dispersion the amount of refraction (index of refraction) diminishes as the length of wave grows shorter. Goldhammer therefore concludes that *a* and *c* can be thus explained by anomalous refraction and dispersion, together with the hypothesis that the X rays are ordinary transverse vibrations of the ether, such as constitute ordinary ultra-violet light. The wave lengths of the X rays is, however, much smaller than those of any hitherto observed ultra-violet rays.

In 1867 M. Boussinesq presented a paper to the French Academy on the Théorie nouvelle des ondes lumineuses,* in which the effects of the momentum communicated to the molecules of matter by the ether are considered to be the cause of reflection, refraction, polarization, dispersion, etc. The ether is supposed to be homogeneous and of the same density and rigidity in all bodies, and that when light enters a transparent medium the molecules of that medium may be set in motion isochronously with the motion of the ether. Sellmeyer also, in 1872, adopted the hypothesis that the ponderable atoms vibrate, but with much smaller amplitudes than the ether particles. The theories of Boussinesq and Sellmeyer lead to expressions for indices of refraction in cases of anomalous refraction which bear upon the X-ray phenomena. The electrical stress acting on the ether may probably serve to set the molecules of the fluorescent substances into their peculiar rates of vibration.

---

* Glazebrook, Optical Theories, British Association Report, 1885.

CHAPTER XXI.

THE SUN.

In asking ourselves What is electricity? we are naturally led to inquire into the constitution of the sun, to which we owe our electrical energy. What, therefore, is the constitution of the sun? It seems strangely analogous to an enormous electrical furnace. If one gazes into an electrical furnace in which there is a mass of molten metal—silver, for instance—one perceives vapours shifting over the glistening mass. In this furnace carbon becomes freed from its impurities; the iron and sodium, for instance, are driven off in vapour, and the pure carbon lies in the heart of the furnace, surrounded by clouds of what once existed throughout its mass.

When one gazes at the spectrum of the sun one marvels at the mysterious arrangement of the dark lines which indicate the absorption of the vapour of some metal. According to the electro-magnetic theory of light, these dark lines represent an absorption of electric energy also. To what is due the electric energy which reaches us in electro-magnetic waves propagated through the infinite space between us and the farthest star?

The spectrum of the sun is like some ancient palimpsest, with inscription upon inscription laid upon each other. A photograph of it is a composite photograph made up of the effects of the vapour of iron and calcium, of cobalt and nickel, of sodium, and many other metals—of possibly all the metals which we know upon this earth. If we could remove one by one the spectrum of these metals, one could obtain the spectrum of the glowing furnace beneath the atmosphere made by the vapours. There seems to be no doubt that certain of the peculiar bands due to carbon can be detected in the solar spectrum. They are, however, almost obliterated by the overlying absorption lines of other metals, especially by the lines due to iron. In order to form an idea of the amount of iron in the vapour of the sun, this amount of iron having an important bearing upon our ideas of the electro-magnetic condition of the sun, I have endeavoured to ascertain how much of the vapour of iron in conjunction with the vapour of carbon would obliterate the banded spectrum of the latter in the atmosphere of the sun. To this end I obtained carbon terminals containing definite proportions of iron and carbon. The iron reduced by hydrogen was distributed uniformly throughout the mass of carbon. Chemical analysis showed that the mixture, so to speak, was homogeneous. Specimens taken from different portions of the carbons showed in the carbons which I burned in the electric arc twenty-eight per cent of iron and seventy-two per cent of carbon.

The method of experimenting was as follows: That portion of the spectrum of the sun which contains traces of this peculiar carbon band lying at wave length 3883·7, which had been almost obliterated by the accompanying lines of absorption of other metals, among them those of iron, was photographed. The pure carbon-banded spectrum was photographed on the same plate immediately below the solar spectrum, and the spectrum of the mixture of iron and carbon immediately below this. It was seen that from twenty-eight to thirty per cent of iron in combination with seventy-two to seventy per cent of carbon almost completely obliterated the peculiar banded spectrum of carbon. This proportion, therefore, of iron in the atmosphere of the sun, were there no other vapours of metals present, would be sufficient to prevent our seeing the full spectrum of carbon.

The light of the sun and that of the electric furnace closely resemble each other. The light of the electric furnace is due to the combustion of carbon. Can we, therefore, conclude that the sun is a vast electric furnace?

An atmosphere of oxygen greatly augments the vividness of the latter. The question, therefore, whether oxygen exists in the sun is closely related to questions in regard to the presence of carbon when we consider the temperature and light of the sun.

If suppositions also are made in regard to the magnetic condition of the atmosphere of the sun, it is of great interest to determine whether oxygen exists there; for oxygen has been shown by Faraday, and later by Prof. Dewar, to be strongly magnetic.

Prof. Henry Draper brought forward evidence to prove the existence of bright oxygen lines in the solar spectrum. Prof. Hutchins, of Bowdoin College, and myself examined this evidence, and, after a long study of the oxygen spectrum in comparison with the solar spectrum, came to the conclusion that the bright lines of oxygen could not be distinguished in the solar spectrum. Since we published our paper, in 1885, I have lately studied the subject from another standpoint. I carefully examined the regions in the solar spectrum where the bright lines of oxygen should occur if they manifest themselves, in order to see if any of the fine absorption lines of iron in the spectrum of iron were absent; for it is reasonable to suppose that the bright nebulous lines of oxygen would obliterate the faintest lines of iron.

The method adopted by Draper for obtaining the spectrum of oxygen consisted in the employment of a powerful spark in ordinary air. To obtain this spark the current from a dynamo running through the primary of a Ruhmkorff coil was suitably interrupted. By the use of an alternating machine and a step-up transformer suitable sparks can be more readily obtained, since the time of exposure with a grating of large dispersion is long. Considerable heat is developed in the transformer from the powerful currents which are necessary to produce a spark of sufficient brilliancy. I have therefore modified the method in the following manner: The spark gap is inclosed in a suitable chamber which can be exhausted; when the exhaustion is pushed to a certain point the length of the spark can be increased ten or twelve times over its length in air, and a suitable spark for photographic purposes can therefore be obtained by the employment of far less electrical energy in the transformer. A pressure of eight to ten inches of mercury in the exhausted vessel is sufficient. A quartz lens inserted in the walls of the exhausted chamber serves to focus the light of the spark on the slit of the spectroscope. A careful examination of the solar spectrum showed that none of even the finest iron lines were obliterated in the spaces where the bright oxygen lines should occur.

Lord Salisbury, in his address before the British

Association at Oxford, 1894, remarks: "Oxygen constitutes the largest portion of the solid and liquid substances of our planet, so far as we know it; and nitrogen is very far the predominant constituent of our atmosphere. If the earth is a detached bit, whirled off the mass, leaving the sun, we cleaned him out so completely of his nitrogen and oxygen that not a trace of these gases remain behind to be discovered even by the sensitive vision of the spectroscope."

Although we have not succeeded in detecting oxygen in the sun, it seems to me that the character of its light, the fact of the combustion of carbon in its mass, the conditions for the incandescence of the oxides of magnesium, of lanthanum, and of the other oxides of the rare earths which exist, would prevent the detection of oxygen in its uncombined state. Notwithstanding the negative evidence which I have brought forward, I can not help feeling strongly that oxygen is present in the sun, and that the sun's light is due to carbon burning in an atmosphere of oxygen.

Can not also the dark spots on the sun be explained by Kirchhoff and Stefan's law, that in a heated space a bundle of rays made up of direct and reflected rays from a surface show the same peculiarities that a bundle of rays from a dark hot body would show? A dark spot on the sun is merely a hole in the gaseous envelope through which we look into an oven. The direct and reflected rays with the supposable layers of vapours of different light-emission power make the interior of the oven of varying light intensity. From the balancing of the reflected and direct rays, a smaller amount of rays reach the eye from certain vapours than from the outer envelope on which this balancing of direct and reflected rays does not take place.

Kirchhoff and Stefan's law has lately received a great deal of attention from experimenters in Germany. W. Wien and O. Lummer suggest that two pieces of thin platinum foil, brought to incandescence by an electric current, be placed opposite each other in a furnace. One is provided with a slit through which the other can be viewed. The inner appears much brighter than the outer, since it is shielded from the reflection of the walls of the oven by the outer foil. The temperature of the pieces of foil can be determined by the increase of resistance of the platinum. The arrangement can be thus used as a bolometer, the radiation to be measured being sent through the slit, and both pieces of platinum being thus heated. In this way a result is obtained which is independent of the individual peculiarities of the absorbing and emitting surfaces, and the absolute radiation can be measured more correctly than by previous methods. Mr. St. John, late holder of the Tyndall scholarship of Harvard University, working with Prof. Warburg in Berlin, also independently of Wien and Lummer, worked out a practical method of measuring the light emitted by various substances at high temperatures. His method suggests the application of Kirchhoff's law to the question of the light of the dark spots on the sun. Two pieces of platinum foil were suspended side by side in an oven which was heated to a high temperature. Observations of their incandescence were made through a suitably placed window by means of a photometer. When one piece of foil was coated with an oxide of the rare earth's zirconium, lanthanum, etc., it appeared brighter than the uncoated piece. If one piece of foil was inclined to the wall of the oven so that the reflected rays from the walls of the oven were sent through the window, the two pieces of foil could be made to appear of the same intensity. The sum of the direct and reflected light is then equal for both pieces of foil. The uncoated piece must reflect just as much more light than the coated as it is deficient in the amount of direct light it can transmit. This is in accordance with Kirchhoff's law—that is, that in a heated space a bundle of rays made up of direct and reflected rays from a surface show the same peculiarities that a bundle of rays from a dark hot body would show. Mr. St. John utilized this idea by bringing a cold porcelain cylinder into the neighbourhood of the pieces of foil; the bare platinum could then be quickly distinguished from the surrounding hot walls, and appeared darker than the coated platinum. As soon as the rod took the temperature of the oven the field of view appeared uniformly bright.

Such are some of the considerations to which I am led by a study of the electric oven; they must be regarded not as final considerations, but merely as attempts to penetrate into the mysteries of the sun.

CHAPTER XXII.

WHAT IS ELECTRICITY?

THE old fluid theories of electricity can be said to have been buried in a common grave with the theories of caloric and phlogiston, and Faraday's researches give the deathblow to the old theory of action at a distance (J. J. Thomson). I have endeavoured to show in the preceding chapters that in the phenomena of the transformations of energy there is a great field of investigation which will repay the student to study and to work in. If we can not discover what electricity is, we can ascertain its relations to light and heat. The human mind, however, is far-reaching, and loves to frame hypotheses and theories, and perhaps no subject is fuller of theories than that of electricity.

If we abandon Maxwell's electro-magnetic theory of light, we find that we must choose between a great number of rival theories—the theories of Ampère, Grassman, Stefan, Korteweg, Neumann, Gauss, Keber, Riemann, and Clausius—in which the actions in a medium between magnets and electrical circuits are not considered. Moreover, the various rotational phenomena of magnetism and electricity are not fully embraced in these theories. These theories, moreover, do not consistently connect the manifestations of light, heat, and electricity, and most of them are not founded on the doctrine of the conservation of energy. Prof. J. J. Thomson, in a report on electrical theories,* divides the theories into the following classes:

"1. Theories in which the action between elements of currents is deduced by geometrical considerations, combined with assumptions which are not explicitly, at any rate, founded on the principle of the conservation of energy. This class includes the theories of Ampère, Grassman, Stefan, and Korteweg.

"2. Theories which explain the action of currents by assuming that the forces between electrified bodies depend upon the velocities and accelerations of the bodies. This class includes the theories of Gauss, Weber, Riemann, and Clausius.

"3. Theories which are based upon dynamical considerations, but which neglect the action of the dielectric. This class contains L. E. Neumann's potential theory and Helmholtz's extension of it.

"4. C. Neumann's theory.

"5. Theories which are based upon dynamical considerations, and which take into account the action of the dielectric. This class includes the theories of Maxwell and Helmholtz."

Prof. Thomson criticises the various theories, and shows "that they can be divided into two great classes, according as they do or do not take into account the action of the dielectric surrounding the various conductors in the field." According to the potential theories of L. E. Neumann and Helmholtz, in an unclosed circuit there are forces which arise from the discontinuity at the ends of the circuit. Shiller's † experiments

* Report of the British Association, Aberdeen, 1885.

† Poggendorf's Annalen, vol. clix, p. 456.

show, however, that this potential theory is wrong "if we neglect the action of the dielectric and assume that the current stops at the end of the circuit." Shiller also showed* that Ampère's theory fails for open circuits, and that Grassman's and Clausius's theories must be wrong as well as Ampère's and Korteweg's, for Shiller's experiments proved that the dielectric must be taken into account.

Maxwell's great theory is full of the intimations which we all have in regard to the probable relation between the varied transformations of energy. It rests, however, upon the hypothesis of the existence of displacement currents in a non-conductor, and the existence of these currents has never been satisfactorily shown by experiment.

When a Leyden jar, for instance, is discharged by connecting the outer coating to the inner by a wire, it is supposed by Maxwell that displacement currents of electricity occur in the glass of the jar, which is the dielectric which separates the outer tin-foil coating of the jar from the inner coating. These currents are called displacement currents, to distinguish them from the apparent current in the wire which discharges the jar. The electrical state on the coatings of the jar is supposed to be rapidly displaced to and fro by the oscillations of the jar. The conduction currents in the wire are transformed into heat. While the displacement currents do not manifest this transformation, they are supposed, however, to exert magnetic influences like ordinary conduction currents. They not only are called into existence in the glass or dielectric of the jar, but they also appear in the air surrounding the jar. They

* Poggendorf's Annalen, vol. clix, p. 456.

are instantaneous currents, and depend on the rate of change of the electro-motive force, or difference of potential between the coatings of the Leyden jar, and also upon the substance of the dielectric, whether it be glass, or rubber, or air.

I have said that the existence of these displacement currents has never been proved by experiments which are free from criticism. The most satisfactory investigation is that of Hertz, who apparently showed that displacement currents in a large pile of books, and in large masses of other dielectrics, exerted a magnetic effect upon the electric waves emanating from an oscillator. The position of the little circle of wire (N, Fig. 47) constituting his exploring resonator was influenced by the proximity of a large solid dielectric, such as a block of paraffin, and he attributed the disturbance to Maxwell's displacement currents in this dielectric. Prof. J. J. Thomson remarks: * "The most pressing need in the theory of electro-dynamics seems to be an experimental investigation of the question of the continuity of these dielectric currents. We have experimental proof that they exist (?), but we do not know whether Maxwell's assumption that they always form closed circuits with the other currents is true or not. If Maxwell's assumption should turn out to be true, we should have a complete theory of electrical action."

What shall we therefore answer to the question, What is electricity? Must we reply, *Ignoramus ignorabimus*—(We are ignorant, and we shall remain ignorant)? We have already strong grounds for believing that we live in a medium which conveys to-and-fro or periodic movements to us from the sun, and

---

* Report on Electrical Theories.

that these movements are electro-magnetic, and that all the transformations of light and heat, and indeed the phenomena of life, are due to the electrical energy which comes to us across the vacuum which exists between us and the sun—a vacuum which is pervaded by the ether, and which is a fit medium for the transmission of the electro-magnetic waves.

INDEX.

THE END.

Spotted something that isn't quite right?

Let us know and we'll fix it for future reprints. And as a thank you for your diligence, we'll send you a free Sothis book of your choice.

publisher@sothis.press

This page intentionally left blank

Made in United States
Orlando, FL
05 November 2026

90229006R00194